Doros T. Petasis
EPR Spectroscopy

Also of Interest

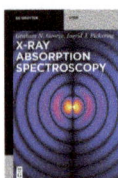

X-ray Absorption Spectroscopy
George, Pickering, 2023
ISBN 978-3-11-057037-3, e-ISBN 978-3-11-057044-1

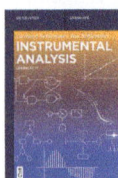

Instrumental Analysis.
Chemical IT
Schlemmer G, Schlemmer J, 2022
ISBN 978-3-11-068964-8, e-ISBN 978-3-11-068966-2

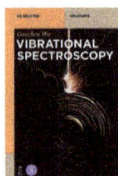

Vibrational Spectroscopy
Wu, 2019
ISBN 978-3-11-066229-0, e-ISBN 978-3-11-066230-6

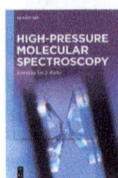

High-pressure Molecular Spectroscopy
Butler (Ed.), 2022
ISBN 978-3-11-066528-4, e-ISBN 978-3-11-066861-2

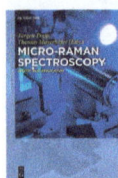

Micro-Raman Spectroscopy.
Theory and Application
Popp, Mayerhöfer (Eds.), 2020
ISBN 978-3-11-051479-7, e-ISBN 978-3-11-051531-2

Doros T. Petasis

EPR Spectroscopy

—

DE GRUYTER

Authors
Prof. Doros T. Petasis
Allegheny College
520 North Main St.
Meadville, PA 16335
United States of America

ISBN 978-3-11-041753-1
e-ISBN (PDF) 978-3-11-041756-2
e-ISBN (EPUB) 978-3-11-042357-0

Library of Congress Control Number: 2022935006

Bibliographic information published by the Deutsche Nationalbibliothek
The Deutsche Nationalbibliothek lists this publication in the Deutsche Nationalbibliografie;
detailed bibliographic data are available on the Internet at http://dnb.dnb.de.

© 2022 Walter de Gruyter GmbH, Berlin/Boston
Cover image: Elerium/E+/gettyimages
Typesetting: Integra Software Services Pvt. Ltd.
Printing and binding: CPI books GmbH, Leck

www.degruyter.com

Preface

This textbook is an introduction to the theory and techniques of Electron Paramagnetic Resonance (EPR). It is intended for advanced undergraduate and beginning graduate students who are first learning about EPR. It is by no means an exhaustive coverage of the field but rather an attempt to introduce a complex topic in a manner that is more accessible to the beginner. There are many excellent books that provide a comprehensive coverage of EPR such as the superb monograph "Electron Paramagnetic Resonance of Transition Series Ions" by Abragam and Bleaney, and classic books such as "Electron Paramagnetic Resonance: Elementary Theory and Practical Applications" by Weil, Bolton and Wertz and "Electron Paramagnetic Resonance: An Introduction to Transition Group Ions in Crystals" by Orton. These books cover a large amount of material that makes them very useful as sourcebooks but can be intimidating to someone first learning about EPR. This textbook follows a development of the material that I myself found useful when first learning about EPR and currently use when teaching EPR to advanced undergraduate students. The first chapter briefly introduces the EPR technique within a historical context followed by a chapter that surveys the quantum mechanical background necessary for understanding EPR theory. Chapter three, the longest chapter in the book, covers the fundamentals of EPR. This chapter presents the necessary information needed to get a very basic understanding of the technique, what EPR spectra look like and how to interpret them. Chapters four and five focus on more specialized topics such as the effect of symmetry on the EPR spectra and the concept of the spin Hamiltonian. The last chapter is focused on experimental techniques including instrumentation details that explain the operation of EPR spectrometers, cryogenic systems necessary for low-temperature experiments and sample preparation.

I would like to acknowledge several people who were instrumental in the creation of this book. Prof. Michael Hendrich of the Department of Chemistry at Carnegie Mellon University was my post-doctoral mentor who introduced me to the applications of EPR in bioinorganic chemistry. During the years I spent in Mike's lab I benefited enormously from his extensive knowledge of all aspects of EPR, which made me a better EPR experimentalist. A few years ago, we co-authored two book chapters on EPR which formed the foundation for this book. Dr. Yiannis Sanakis of the National Center for Scientific Research Demokritos in Athens, Greece, read and provided valuable feedback on most of the chapters. My former students, Prof. Katlyn Meier of the Department of Chemistry at the University of Miami and Prof. Bill Gunderson of the Department of Chemistry at Hendrix College also read and provided corrections and detailed comments on several chapters. I would also like to express my gratitude to Dr. Ria Sengbusch, project editor for chemistry and materials sciences at De Gruyter. Her support, guidance and suggestions through the various stages of the development of this book were invaluable. Finally, I would like to acknowledge the support and encouragement that my wife Leslie and my daughter Emma provided throughout the writing of this book. Without them this book would have never reached completion.

https://doi.org/10.1515/9783110417562-202

Contents

1 Introduction

Electron paramagnetic resonance (EPR or electron spin resonance) is a highly sensitive, nondestructive spectroscopic technique that employs microwave frequencies and moderate magnetic fields in the study of systems incorporating paramagnetic species. Such species contain unpaired electrons, giving them nonzero magnetic moment due to the intrinsic spin of the electron. They include transition series and rare earth ions, organic free radicals and inorganic molecules with unpaired electrons (such as O_2 and NO). EPR is highly selective, specifically targeting the paramagnetic entities within the sample. It provides a wealth of information about their properties, such as the identity of the paramagnetic species, its oxidation and spin state and interactions with the surrounding ligands. In addition, EPR provides information on the nature and strength of interactions between paramagnetic species within the same molecule or crystal.

EPR is highly versatile in the nature of samples that it can study. Solid-state, liquid, gas and biological samples are all accessible to EPR. It is particularly suited for the study of noncrystalline samples, which makes it a good complementary technique to X-ray diffraction in providing structural information. The versatility and specificity of EPR makes it a good complement to other techniques as well, such as X-ray absorption spectroscopy (XAS), magnetic circular dichroism (MCD), magnetic susceptibility and Mössbauer spectroscopy in the study of a very broad range of systems [1].

The discussion in this chapter aims to outline the EPR phenomenon and establish its nature using some elementary physics. Subsequent chapters will delve deeper into the nature of EPR and address some advanced topics to outline fully the details of EPR. Before discussing the EPR phenomenon, a brief history of the discovery and development of EPR will be presented.

1.1 Brief historical overview

Continuous-wave EPR (cw-EPR) was discovered by Evgeny Zavoisky at Kazan State University in 1944 when he first detected resonance absorption in the salts of manganese sulfate, copper sulfate and chromium chloride [2]. The technique benefited from technological advances that occurred during the development of radar during WWII such as the production of high-power microwave radiation generators, sensitive microwave detectors and amplifiers for signal processing [3]. The invention and commercialization of the klystron microwave generator by the Varian brothers as a stable and consistent source of monochromatic microwave radiation led to the development of the first commercial EPR spectrometer that was produced in the mid-1950s by the Varian Corporation. This spectrometer operated in the microwave region of the

https://doi.org/10.1515/9783110417562-001

electromagnetic spectrum in the frequency range of 8.5–12 GHz, referred to as X-band. This is still the most popular microwave band for EPR spectroscopy since it is quite sensitive, and the microwave components required for spectrometer assembly are inexpensive and easy to find. Varian also developed a series of sophisticated and very successful research-grade spectrometers in the 1960s and 1970s, the E and Century lines of spectrometers, some of which are still in use today as shown in Fig. 1.1.

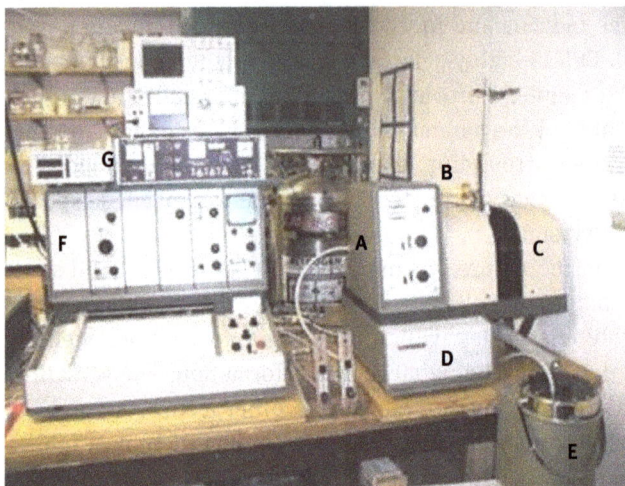

Fig. 1.1: Varian E-3 X-band spectrometer equipped with a liquid nitrogen cryostat in the Department of Physics at Allegheny College. The E-3 line of instruments was the first commercially available line of research grade spectrometers. The robustness of these instruments is demonstrated by the fact that they are still in operation today, primarily as teaching instruments. A, microwave bridge; B, microwave probe; C, electromagnet; D, magnet power supply; E, liquid nitrogen cryostat; F, control console with plotter; G, temperature controller.

EPR was widely applied in the area of solid-state physics during the 1950s, especially in the Laboratory of Bleaney at Oxford University where the effectiveness of the technique as an analytical instrument was demonstrated in the groundbreaking studies of the salts of transition series and rare earth (lanthanide) elements such as ethylsulfates and double nitrates [4]. An early EPR spectrum of a crystal of Nd(III) ethylsulfates obtained by Bleaney's group is shown in Fig. 1.2. A large number of papers followed on the research carried out by such EPR pioneers as Baker, Elliott, Stevens and many others [5–8]. The versatility of EPR in solid-state physics was further demonstrated in the application of the technique to the study of color centers in crystals by Hutchison [9] at the University of Chicago with the spectra successfully interpreted by Kittel's group at Berkeley [10]. The need to determine the nature of color centers produced by radiation damage in crystals led to the development of sensitive home-built spectrometers, in particular in Seitz's laboratory at the University of Illinois [11].

Extensive applications in chemistry followed, centered on the study of free radicals. Biological systems were first studied in the late 1950s, when the first EPR signals from single crystals of myoglobin were recorded by Ingram and Kendrew [12]. In the last few decades, EPR found applications in biophysics and bioinorganic chemistry, in combination with other techniques (such as Mössbauer, MCD and XAS), and has proved its utility in the study of biological and synthetic systems. The versatility of EPR in bioinorganic chemistry is underscored by the fact that it can be applied to both powder (including frozen solution samples) and crystalline systems, which is important in the study of biological molecules where many systems cannot be crystallized. In addition, along with free radicals, EPR also targets metal ions with unpaired electrons making it ideal for the study of the active sites of metalloproteins, one of the largest areas of bioinorganic chemistry. All of the essential biochemical processes of metalloproteins occur at the active site, where the metal ion(s) are bound, making EPR an indispensable tool in understanding the function of these very important molecules.

(a)   (b)

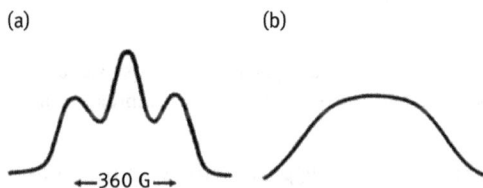

Fig. 1.2: One of the early EPR spectra of Nd(III) ethylsulfate crystal at 20 K obtained by Bleaney, Elliott and Scovil. The magnetic field is oriented (a) parallel and (b) perpendicular to the hexagonal axis of symmetry of the crystal. The triplet structure observed in (a) features symmetric satellite lines relative to the central line that are due to dipole–dipole interactions between nearest neighbor Nd(III) ions. These lines are broadened and not resolved in (b) [13].

EPR spectroscopy has continued to evolve with the development of spectrometers that operate at frequencies other than X-band (8–12 GHz). Lower frequencies such as L-band (1–2 GHz) and S-band (2–4 GHz) allow better resolution of hyperfine lines (hyperfine lines are discussed in Chapter 3). Higher frequencies such as K-band (18–26 GHz) and Q-band (26–40 GHz) allow better signal-to-noise ratios. In recent years, high-frequency EPR has been extended to W-band (75–110 GHz) and beyond to reach quasi-optical frequencies (>250 GHz). A number of associated techniques have also been developed such as electron nuclear double resonance developed by Feher in the late 1950s [14] and pulsed techniques such as electron spin echo envelope modulation. The technique has even extended to in vivo EPR that allows the imaging of animal tissue and organs [15]. Discussions of these topics can be found in many chapters and specialized books. This book is primarily focused on discussions of the fundamental principles of cw-EPR with brief overviews of some associated techniques.

EPR spectrometers continued to advance in recent decades by incorporating new detection techniques, state-of-the-art resonators with very high quality factors

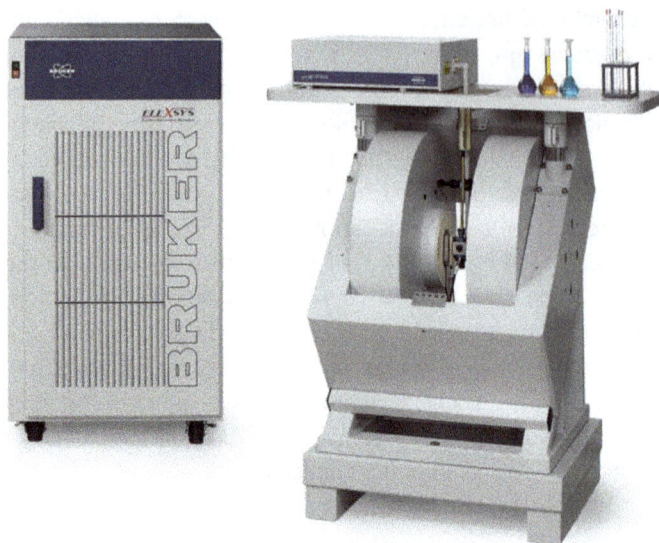

Fig. 1.3: Bruker's Elexsys E500 cw-EPR X-band spectrometer that utilizes a multipurpose signal-processing unit, an ultra-low-noise dual Gunn source, and super-high-Q cavity (www.bruker.com).

(Q-factor), microwave sources such as the very stable/powerful Gunn diode and computational capabilities. Presently, two commercial companies dominate the market for research-grade EPR spectrometers: Bruker with its Elexsys series of spectrometers (Fig. 1.3) and JEOL with its JES-FA series of spectrometers. Special-purpose instruments are continually being developed by individual research groups such as the Q-band spectrometer in the laboratory of Prof. Michael Hendrich of the Department of Chemistry at Carnegie Mellon University. This instrument utilizes a novel probe and resonator design making it capable of detecting signals from both half-integer and integer-spin systems at low temperatures that allow the observation of EPR signals from integer-spin complexes that were previously thought to be EPR silent [16]. Spectra obtained with this spectrometer from a sample of cytochrome c554 are shown in Fig. 1.4, which demonstrates the utility of multimode, multifrequency EPR in studying systems with complex spectra.

1.2 Basic principles of EPR

1.2.1 Atomic magnetic moments

Paramagnetic species contain unpaired electrons that possess a net magnetic moment $\vec{\mu}$ that allows them to interact with an externally applied magnetic field, such as the one generated by a laboratory electromagnet. Classically, the simplest magnetic moment is created when a current I is flowing through a wire in the shape of a

(a)

(b)

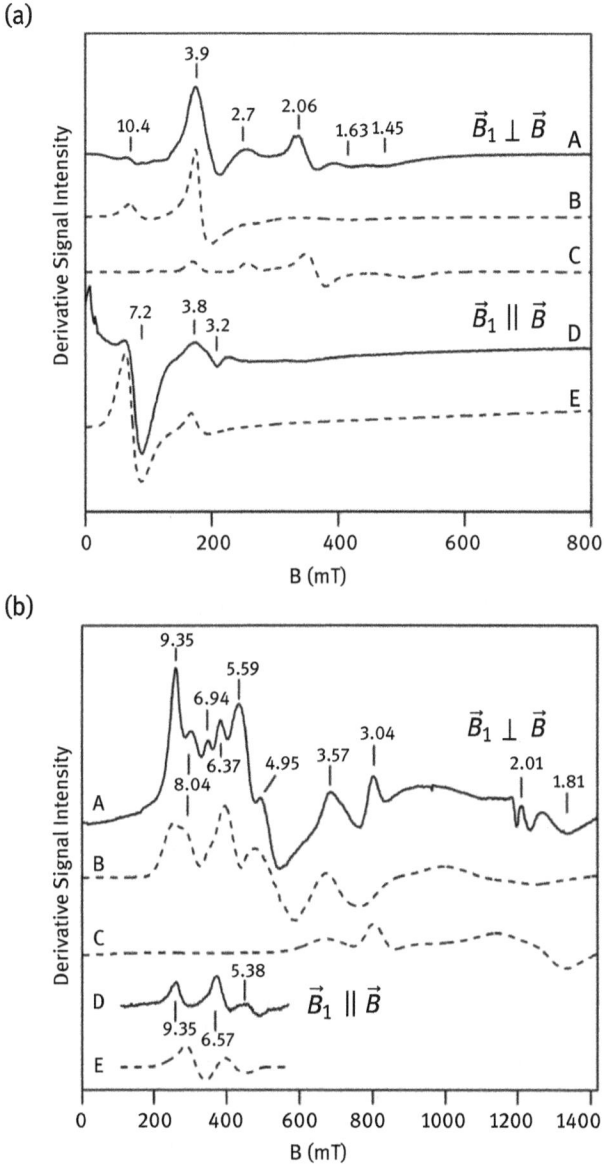

Fig. 1.4: The utility and sophistication of modern EPR is demonstrated in this example of multifrequency, multimode EPR: experimental spectra (solid lines) are taken at two different frequencies ((a) X-band and (b) Q-band) with two different orientations of the microwave magnetic field ($\vec{B}_1$) relative to the external magnetic field ($\vec{B}$) allowing the detection of signals from both half-integer and integer-spin species. Simultaneous quantitative simulations (dashed lines) that fit the data across both frequencies allow analysis of complex spectra. The sample is oxidized cytochrome c_{554}, a tetraheme cytochrome from *Nitrosomonas europaea*, which is an essential electron transfer component in the biological oxidation of ammonia [17].

circular loop of radius r. This current flow gives rise to a magnetic moment $\vec{\mu}$ with magnitude equal to $\mu = I\pi r^2$, passing through the center of the loop with a direction perpendicular to the plane of the loop as determined by the right-hand rule (Fig. 1.5 (a)). This magnetic moment can interact with an external magnetic field $\vec{B}$ with a potential energy of interaction between $\vec{\mu}$ and $\vec{B}$ given by

$$E = -\vec{\mu} \circ \vec{B} = -\mu B \cos\theta \tag{1.1}$$

where θ is the angle between $\vec{\mu}$ and $\vec{B}$. The minimum energy state occurs when $\vec{\mu}$ is parallel to $\vec{B}$ (Fig. 1.5 (b)).

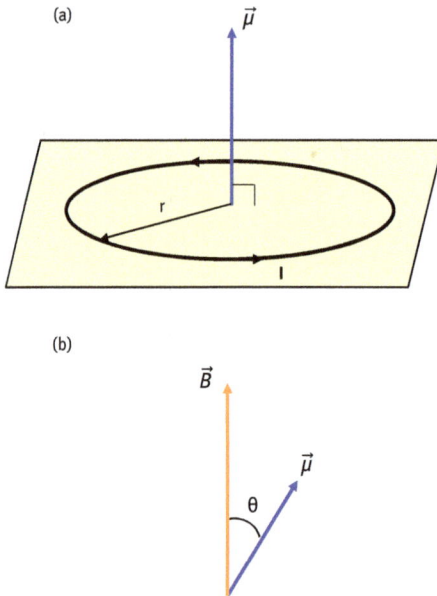

Fig. 1.5: (a) A magnetic dipole moment is generated by a current I flowing around a circular loop. $\vec{\mu}$ is perpendicular to the plane of the loop passing through the center of the loop with a direction given by the right-hand rule. $|\vec{\mu}| = IA$, where A is the area enclosed by the current loop. (b) A magnetic moment inside an external magnetic field $\vec{B}$ at an arbitrary angle θ.

The electron has two types of magnetic moments, which are connected to its motion inside the atom. These two magnetic moments are associated with the angular momenta that result from the motion of the electron: an orbital angular momentum due to the motion around the nucleus and an intrinsic "spin" angular momentum. To understand the origin of these magnetic moments, we have to consider a mechanical model of the motion where we think of the electron as a particle of mass m_e ($m_e = 9.10938 \times 10^{-31}$ kg) moving with velocity $\vec{v}$ in a circular trajectory of radius r and centered at the nucleus at O. According to classical mechanics, the electron will have an angular momentum $\vec{L}$ that is perpendicular to the plane of the trajectory passing through O, with magnitude

$$|\vec{L}| = m_e v r \tag{1.2}$$

as shown in Fig. 1.6. A magnetic dipole moment $\vec{\mu}_L$ is associated with this motion when we also consider that the electron is a charged particle with charge $-e$ ($e = 1.60218 \times 10^{-19}$ C). The circular motion of such a charge gives rise to a circulating current I in the opposite direction to the motion of the electron forming a current loop that overlaps with the orbit of the electron. This generates a magnetic moment $\vec{\mu}_L$ with a direction opposite to that of $\vec{L}$ and perpendicular to the plane of the loop. The magnetic moment and the angular momentum are related according to the expression

$$\vec{\mu}_L = -\frac{e}{2m_e}\vec{L} \tag{1.3}$$

where e is the electronic charge and m_e the mass of the electron as defined earlier. In a similar manner, we can think of the electron possessing an intrinsic angular momentum $\vec{S}$ with an associated magnetic moment $\vec{\mu}_S$ as shown in Fig. 1.6. The relationship between the two is given by

$$\vec{\mu}_S = -\frac{e}{m_e}\vec{S} \tag{1.4}$$

This is the magnetic moment of the electron that interacts with an external magnetic field, whereas the contribution of the orbital angular momentum appears through the spin–orbit interaction, a mechanism that allows the two magnetic moments $\vec{\mu}_L$ and $\vec{\mu}_S$ to interact and will be discussed in more detail in the next chapter.

We saw earlier that there is a potential energy present when a magnetic moment is placed inside an external magnetic field. In addition to this energy of interaction between $\vec{\mu}$ and $\vec{B}$, $\vec{\mu}$ also experiences a torque

$$\vec{\tau} = \vec{\mu} \times \vec{B} \tag{1.5}$$

that tends to align $\vec{\mu}$ with $\vec{B}$ to minimize the energy of interaction [18]. If there are no pathways for transferring energy away from the system, the potential energy of interaction will not change and instead $\vec{\mu}$ will be forced to precess about $\vec{B}$ at a constant angle θ resulting in a trajectory that lies on the surface of an inverted cone as shown in Fig. 1.7. For the case of the electron where the magnetic moment is associated with an angular momentum vector, this phenomenon is called Larmor precession and will be discussed in more detail in later chapters. The frequency of precession is given by

$$\vec{\omega}_l = \frac{e}{2m_e}\vec{B} \tag{1.6}$$

(a)

(b)

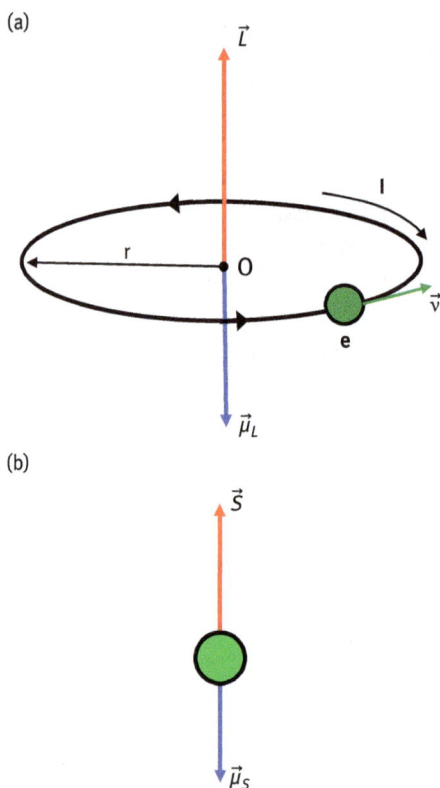

Fig. 1.6: (a) The simplest model of the orbital angular momentum $\vec{L}$ of the electron is the one where the electron moves about the nucleus in a circular trajectory as shown here. $\vec{L}$ passes through the center of the circular trajectory perpendicular to the plane of the orbit. This motion of the electron also generates a current with conventional direction in the opposite sense as that of the electron motion. The result is a circular current loop generating a magnetic moment $\vec{\mu}_L$ in the opposite direction to that of $\vec{L}$. (b) The spin angular momentum $\vec{S}$ with its associated magnetic moment $\vec{\mu}_S$ can be classically illustrated in a similar way by thinking of the charged electron as "spinning" about its axis of symmetry. This is, of course, a very naïve analogy since $\vec{S}$ is a quantum mechanical quantity with no classical analogue. It is predicted by the Dirac equation, which is a result of relativistic quantum mechanics.

and called the Larmor frequency [19]. If the magnetic field is applied along the z-axis, the electronic magnetic moment will then be constrained to move on the surface of an inverted cone with its axis of symmetry along the z-axis. In this case, the magnetic moment can be resolved into two components, $\mu_{||}$ along the z-axis and $\mu_\perp$ in the xy-plane as shown in Fig. 1.7. The $\mu_{||}$ component remains constant in time but the $\mu_\perp$ rotates in the xy-plane at the Larmor frequency. Consequently, we can interact with the electronic magnetic moment via its $\mu_\perp$ component by introducing a rotating magnetic field

$$\vec{B}_\perp = \vec{B}_1 \cos \omega t \qquad (1.7)$$

in the xy-plane. The introduction of this field produces a torque $\vec{\mu} \times \vec{B}_\perp$ acting on $\vec{\mu}$ which tends to change the angle θ. Resonance will occur when $\vec{B}_\perp$ rotates in-phase with $\vec{\mu}$ at the Larmor frequency causing $\vec{\mu}$ to either move toward the z-axis, with a concomitant emission of energy, or move away from the z-axis with absorption of energy.

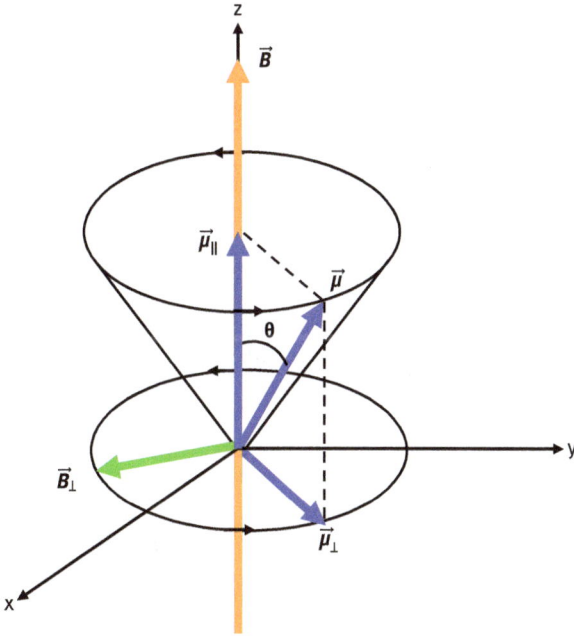

Fig. 1.7: Precession of a magnetic moment $\vec{\mu}$ about an external magnetic field $\vec{B}$ directed along the z-axis. The projection of $\vec{\mu}$ along the z-axis ($\vec{\mu}_\parallel$) is constant but the projection on the xy-plane ($\vec{\mu}_\perp$) is rotating about the z-axis at the Larmor frequency. Resonance is achieved when an external time-varying magnetic field applied in the xy-plane and oscillating at the Larmor frequency interacts with the rotating component of $\vec{\mu}$.

This process provides a classical description of the EPR phenomenon: when a system with an unpaired electron is placed inside an external magnetic field $\vec{B}$, the magnetic moment of the electron $\vec{\mu}_S$ will precess about $\vec{B}$ in the manner illustrated in Fig. 1.7, with $\vec{\mu}$ replaced by $\vec{\mu}_S$. For electronic magnetic moments, the Larmor frequency of precession is in the GHz range. Interaction with the spin magnetic moment results when the rotating magnetic field $\vec{B}_\perp$ is an oscillating electromagnetic field with a frequency of oscillation in the microwave (GHz) range, applied orthogonally to the external field $\vec{B}$. EPR spectrometers are configured in such a way to maintain the condition $\vec{B} \perp \vec{B}_\perp$, which is referred to as "perpendicular-mode" EPR. Fig. 1.8 demonstrates how

(a)

(b)

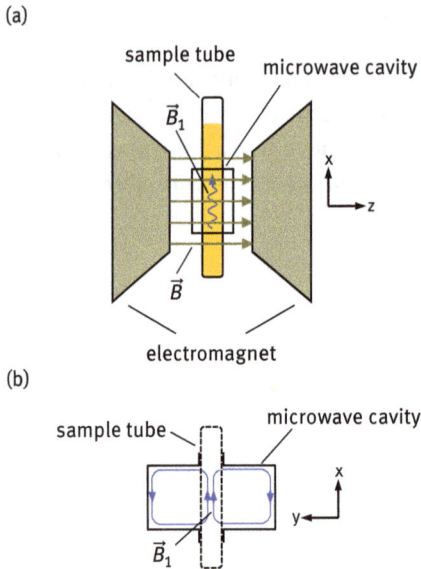

Fig. 1.8: Relative orientation of $\vec{B}$ and $\vec{B}_1$ during an EPR experiment: (a) front view of cavity with $\vec{B}_1$ inside the electromagnet and (b) side view of rectangular resonant cavity in TE$_{102}$ mode showing $\vec{B}_1$ inside the cavity; and $\vec{B}$ is perpendicular to the plane of the page.

this is achieved during a typical EPR experiment. "Parallel-mode" EPR also exists and will be discussed at a later point. This phenomenological discussion of EPR will be further expanded in Chapter 3 where connections to relaxation times and lineshapes will also be made.

1.2.2 Quantum mechanical discussion

Of course, this classical explanation of the EPR phenomenon is inadequate, and a full quantum mechanical explanation is necessary. In order to achieve this goal, a very brief introduction to the techniques and calculations of quantum mechanics is necessary. A more detailed presentation of the quantum mechanical background necessary to understand EPR will be given in the next chapter. Quantum mechanically, the electronic magnetic moment can only have two possible orientations relative to B: one where $\mu_{||}$ is along the positive z-axis and one where $\mu_{||}$ is along the negative z-axis. This is due to space quantization which states that the spin angular momentum can only have two possible orientations relative to an axis of quantization defined by the external magnetic field which is typically placed along the z-axis. The projection of S onto the z-axis has magnitude $S_z = \pm (\hbar/2)$ as shown in Fig. 1.9. In addition, the magnitude of S can only have certain values given by

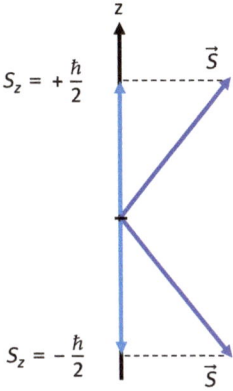

Fig. 1.9: Space quantization: the spin angular momentum $\vec{S}$ has only two allowed orientations relative to the quantization axis z with projections of magnitude $S_z = \pm(\hbar/2)$. The quantization axis is determined by the direction of the external magnetic field.

$$|\vec{S}| = \sqrt{s(s+1)}\hbar \tag{1.8}$$

where $s = \frac{1}{2}$ and $\hbar = (h/2\pi)$. The spin magnetic moment can then be written as

$$\vec{\mu}_S = -\frac{e}{m_e}\vec{S} = -g_e\frac{\mu_B}{\hbar}\vec{S} \tag{1.9}$$

where g_e is the electronic g-value which is equal to ~ 2.0023 for free electrons and

$$\mu_B = \beta = \frac{e\hbar}{2m_e} = 9.274 \times 10^{-24}\frac{J}{T} \tag{1.10}$$

is the Bohr magneton (β is the original symbol for the Bohr magneton used in spectroscopy and it is the one generally used by EPR spectroscopists to this day).

The energy of interaction between $\vec{\mu}_S$ and B is then equal to

$$E = -\vec{\mu}_S \circ \vec{B} = g_e\frac{\beta}{\hbar}\vec{S}\circ\vec{B} = g_e\frac{\beta}{\hbar}S_z B \pm \frac{1}{2}g_e\beta B \tag{1.11}$$

which means that the electron can occupy two possible spin energy states in the presence of an external magnetic field.

Turning on an external magnetic field to a value B_0 will shift the electronic states relative to the zero field value as shown in Fig. 1.10 with an energy separation

$$\Delta E = E_2 - E_1 = g_e\beta B_0 \tag{1.12}$$

This is the Zeeman effect which will be discussed in more detail in a subsequent chapter. At normal temperatures and thermal equilibrium, there will be a larger population of electrons on the lowest state according to the Boltzmann distribution yielding a ratio of populations equal to

$$\frac{N_1}{N_2} = \exp\left(\frac{\Delta E}{k_B T}\right) \qquad (1.13)$$

where k_B is the Boltzmann constant [20].

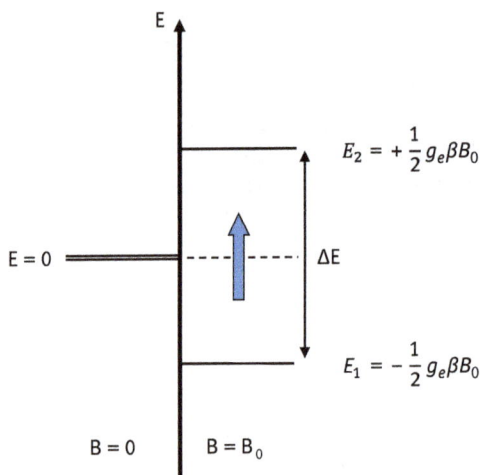

Fig. 1.10: The two possible energy states of a free electron in zero field and a field equal to B_0. The zero-field degenerate state splits into two energy states in the presence of B_0 with energy separation equal to $\Delta E = E_2 - E_1 = g_e \beta B_0$.

Electronic transitions from the ground to the excited state can be induced when electromagnetic radiation of energy $E = h\nu_0 = \Delta E$ shines on the ground state. As long as the inequality $N_1 > N_2$ is maintained, transitions will continue to occur. There are several relaxation mechanisms to prevent the spin populations of the two states from becoming equal (saturation) and thus stopping the absorption of energy. This can happen when energy is transferred from the excited electrons to other parts of the system, which will be discussed later.

1.2.3 The EPR experiment

The frequency of the radiation-inducing transitions between the two electronic energy states is in the microwave region. During an EPR experiment, a microwave generator produces radiation at a fixed frequency while the magnetic field is scanned from a zero field to typically a field of 1 T. Absorption of microwave power occurs at a field $B = B_0$, where the resonance condition $h\nu_0 = g_e \beta B_0$ is satisfied. The microwave power is monitored, and the absorption appears as a bell-shaped curve as shown in Fig. 1.11(a).

This absorption signal is typically sent to a lock-in amplifier which is a phase-sensitive detector that produces a first derivative curve of the absorption as shown in Fig. 1.11(b). This is the characteristic signal obtained from a single free electron that typically exists in free radicals. Such a signal is referred to as an "EPR line." The first derivative presentation of the EPR line has advantages over the direct absorption line since partially overlapping lines in more complicated spectra can be better resolved.

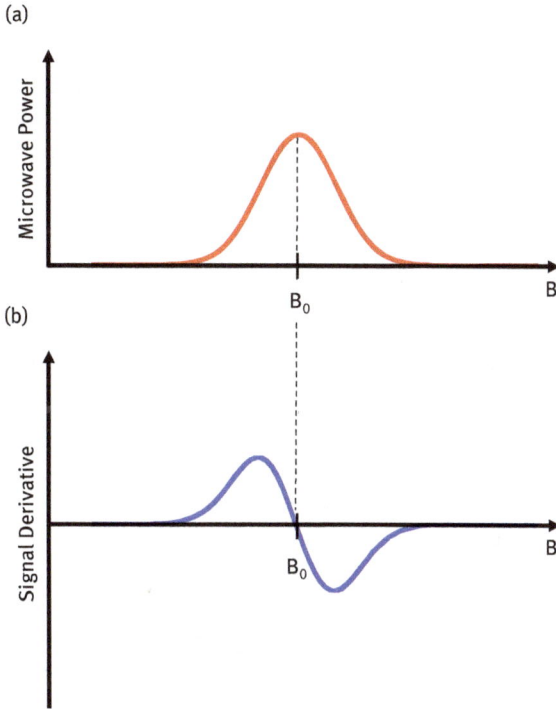

(a)

(b)

Fig. 1.11: A typical absorption line (a) with the first derivative (b). The resonance occurs at a magnetic field B_0 where the absorption line is at a maximum and the first derivative line crosses the baseline.

The most widely used band used by EPR spectrometers is the X-band, which requires dimensions of microwave components in the centimeter range that makes spectrometers easy to construct. Larger frequencies require smaller microwave components but spectrometer technologies have been developed to allow the construction of spectrometers operating at very high frequencies such as Q- and W-bands to allow detection of signals that cannot be detected at lower frequencies ("EPR-silent" samples) and to also improve the resolution of complicated signals. The magnetic fields required for X-band EPR are less than 1 T but higher fields are required for the higher bands. The magnets required for EPR do not have to produce very high fields but are rather required to produce very homogeneous fields at the sample site. Since there are several different microwave bands, the signal from the same sample will appear at different values of the magnetic field B_0 depending on the frequency of the microwave radiation. Consequently, the position of the EPR line is quoted in terms of the g-value which is determined from the resonance condition expressed in terms of the frequency quoted in GHz and the field position B_0 in mT as follows:

$$g = \frac{71.449 \; v \,(\text{GHz})}{B_0 \,(\text{mT})} \tag{1.14}$$

For free electrons $g = 2.0023$, but for electrons bound to atoms and molecules the g-value deviates from this value based on the characteristics of their orbitals. Signals with $g \approx 2.0$ are typical of free radicals which contain electrons on delocalized orbitals.

1.2.4 EPR lineshapes

The EPR lineshape depends on many factors and can yield quite a bit of information about the sample. Firstly, the integrated intensity of the absorption line is a function of the concentration of the sample, the frequency of the microwave radiation (higher frequencies produce more intense signals), the incident microwave power, the transition probability, and the temperature. Proper accounting of these factors allows the integrated intensity to be used to determine the concentration of the spins contributing to the signal. The characteristic bell shape of an EPR line is mathematically described by a Gaussian curve for inhomogeneously broadened lines and a Lorentzian curve[1] for homogeneously broadened lines (homogeneous broadening and inhomogeneous broadening are defined in Chapter 3). The general functions describing these curves are given by

$$f(B) = Ae^{-b^2(B-B_0)^2} \tag{1.15}$$

for a Gaussian and

$$f(B) = \frac{A}{1 + b^2(B - B_0)^2} \tag{1.16}$$

for a Lorentzian where A and b are constants. The signal displayed in Fig. 1.11 is for a Gaussian curve where the inhomogeneous broadening could be due to unresolved hyperfine structure or internal molecular strains. The main characteristic of the EPR line is the full width at half maximum which is called the linewidth ΔB. Transitions

---

[1] In certain cases, the EPR line is a convolution of a Gaussian and a Lorentzian lineshape.

between sharply defined energy levels as displayed in Fig. 1.10 would produce an
EPR line with $\Delta B \rightarrow 0$. The linewidth, however, has a nonzero value due to "lifetime
broadening" which is a result of the very short relaxation times (of the order of 10^{-7} s
or less) [21] that limit the time electrons spend on the excited state E_2. If the lifetime
of the excited state is Δt, then according to the Heisenberg uncertainty principle the
linewidth is $\Delta B \propto 1/\Delta t$. In many systems, Δt can be extremely short at room temperature (e.g., rare earth ions) producing large values for ΔB which result in very broad
lines that make it difficult to detect EPR signals. Such situations necessitate a slowing
down of the transfer of energy from the spin system in order to increase Δt. Various
mechanisms contribute to how fast relaxation times can be with the two most important being spin–lattice and spin–spin relaxation. In spin–lattice relaxation, energy is
transferred from the spin system to the lattice via spin–orbit coupling. Spin–spin relaxation is important in concentrated samples where the energy is transferred through interactions between magnetic moments with dipolar interactions being the dominant
kind of interaction. Both of these mechanisms produce homogeneous broadening and
will be extensively discussed in later chapters.

1.2.5 Basic EPR spectrometer

The simplest spectrometer utilizes a source of electromagnetic radiation, typically
in the visible part of the spectrum, which sends the radiation through a transparent
sample. The transmitted light arrives at a detector, which allows it to be analyzed to
extract information on the optical properties of the sample (Fig. 1.12(a)). EPR spectrometers use microwave radiation which needs to be transmitted through metallic
waveguides as shown in Fig. 1.12(b). In the simplest possible EPR spectrometer, this
radiation travels into a metallic rectangular resonant cavity that is situated between
the pole caps of an electromagnet. The cavity concentrates the radiation and allows
it to interact with the sample. The transmitted radiation reaches a diode detector
that acts as a microwave rectifier, converting microwave power into a current that
can be processed and analyzed. A more practical spectrometer is shown in Fig 1.12(c),
where the detector is not put in line with the source and the cavity. This design allows the cavity to be placed inside a cryostat, which is in turn placed between the
pole caps of an electromagnet. The cryostat can cool the sample to very low temperatures to slow down spin–lattice relaxation times and produce sharp narrow signals.
Most cryostats utilize liquid helium as a coolant and incorporate heaters that allow
them to operate over a large variable temperature range, from around 2 K to room
temperature. Experiments can also be done at temperatures above room temperature
where, instead of a cryostat, an oven can be used.

It is apparent from this discussion that cw-EPR spectrometers are quite complex. This will become more obvious when we discuss EPR instrumentation in
Chapter 6. Even in the early days of EPR, sensitive instruments required a high level

(a)

Source   Sample   Detector

(b) Waveguide

Microwave   Detector
Source
Sample in
Cavity
                Electromagnet

(c)   Circulator

                                Cryostat

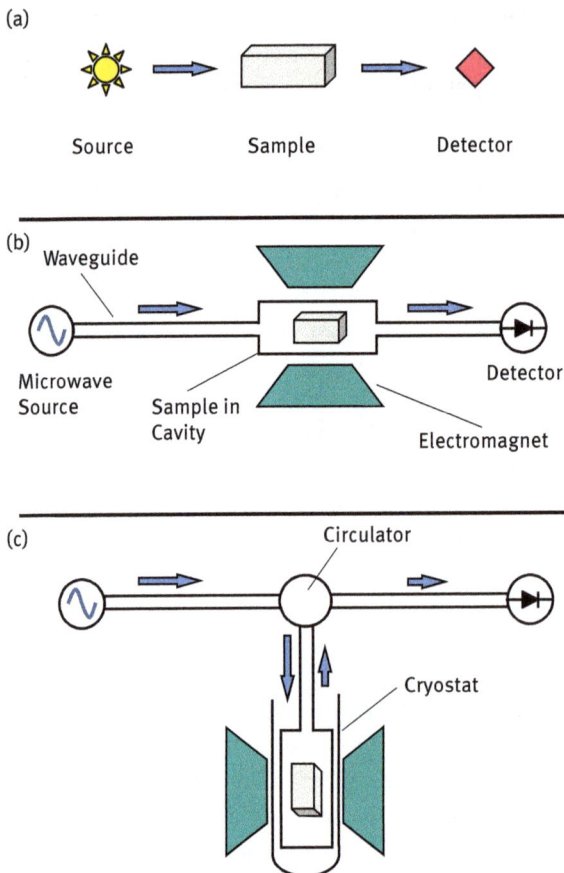

Fig. 1.12: Schematic diagrams outlining the basic principles of operation for a simple optical spectrometer (a) and EPR spectrometers (b–c). The blue arrows indicate the direction of propagation of the electromagnetic radiation. The arrangement in (c) shows the use of an immersion cryostat that allows measurements to be carried out at liquid nitrogen and helium temperatures. A variant of this method utilizes a continuous-flow cryostat used by most modern EPR spectrometers.

of sophistication. Most of the needs of EPR experimenters today are satisfied by commercial instruments but homebuilt spectrometers can be found in labs that require very specific information from samples beyond identifying the paramagnetic species, its valence, and spin states. One such example is the spectrometer mentioned at the end of Section 1.1 and shown in Fig. 1.13. This instrument incorporates a commercial microwave bridge, immersion cryostat, and rotatable electromagnet with a homebuilt microwave probe and cavity. The instrument operates at Q-band frequencies and allows measurements of integer–spin systems with large zero-field splittings, as mentioned in Section 1.1 [16].

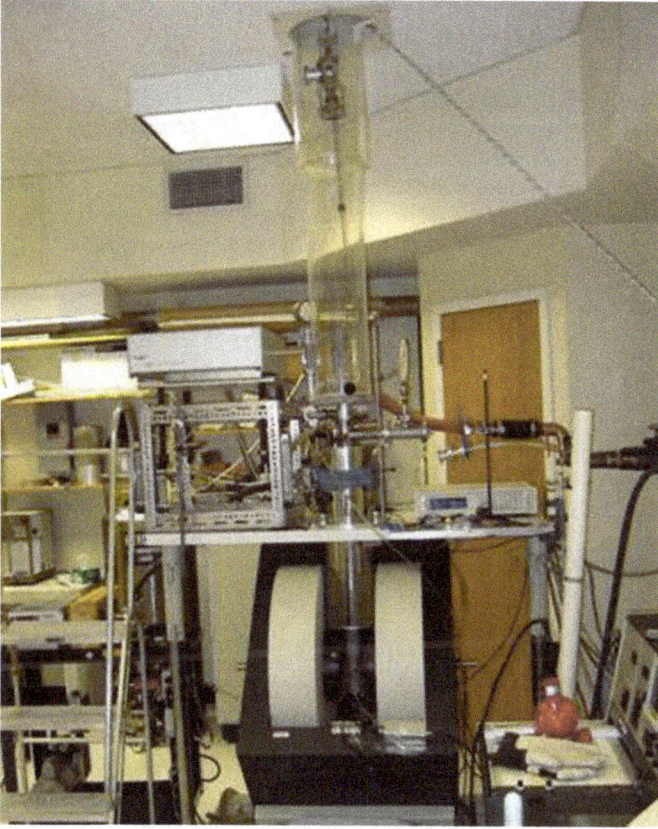

Fig. 1.13: Q-band EPR spectrometer that utilizes commercially available microwave bridge and console (Bruker), a custom-built liquid-helium immersion cryostat (CRYO Industries of America), and a rotatable electromagnet along with a homebuilt EPR probe. This probe is coupled to a cylindrical cavity via a sidewall that allows a parallel orientation of $\vec{B}_\perp$ and $\vec{B}$ for the study of integer-spin systems. (Hendrich lab, Department of Chemistry, Carnegie Mellon University).

Works cited

[1] K. Sauer, Methods in Enzymology-Biochemical Spectroscopy, vol. 246, K. Sauer, Ed., San Diego: Academic Press, Inc., 1995.
[2] G. Eaton, S. Eaton and K. Salikhov, Foundations of Modern EPR, World Scientific Publishing Company, 1997.
[3] C. P. Poole, Electron Spin Resonance: A Comprehensive Treatise on Experimental Techniques (2nd ed.), Dover, 1996.
[4] A. Abragam and B. Bleaney, Electron Paramagnetic Resonance of Transition Ions, Oxford University Press, 1970.
[5] R. J. Elliott and K. W. H. Stevens, "The Theory of the Magnetic Properties of Rare Earth Salts: Cerium Ethyl Sulphate," *Proceedings of the Royal Society A*, vol. 215, p. 437, 1952.

[6]   B. Bleaney and H. E. D. Scovil, "Paramagnetic Resonance in Praseodymium Ethylsulphate," *Philosophical Magazine*, vol. 43, p. 999, 1952.

[7]   R. J. Elliott and K. W. H. Stevens, "The Paramagnetism of Cerium Ethylsulphate: Theory," *Proceedings of the Physical Society A*, vol. 64, p. 932, 1951.

[8]   J. M. Baker and B. Bleaney, "Paramagnetic Resonance in some Lanthanon Ethyl Sulphates," *Proceedings of the Royal Society A*, vol. 245, p. 156, 1958.

[9]   C. A. Hutchison and G. A. Noble, "Paramagnetic Resonance Absorption in Additively Colored Crystals of Alkali Halides," *Physical Review*, vol. 87, pp. 1125–1126, 1952.

[10] A. Kip, C. Kittel, R. A. Levy and A. M. Portis, "Electronic Structure of F-centers. Hyperfine Interaction in Electron Spin Resonance," *Physical Review*, vol. 91, pp. 1066–1071, 1953.

[11] J. Teichmann and K. Szymborski, "Point Defects and Ionic Crystals: Color Centers as the Key to Imperfections," in Out of the Crystal Maze, Oxford, Oxford University Press, 1992, pp. 236–316.

[12] D. J. E. Ingram and J. C. Kendrew, "Electron Spin Resonance in Myoglobin and Hæmoglobin: Orientation of the Hæm Group in Myoglobin and its Relation to the Polypeptide Chain Direction," *Nature*, vol. 178, pp. 905–906, 1956.

[13] B. Bleaney, R. J. Elliott and H. E. D. Scovil, "Dipole-Dipole Interaction in the Rare-Earth Ethylsulphates," Proceedings of the Physical Society A, vol. 64, p. 933, 1951.

[14] G. Feher, "Observation of Nuclear Magnetic Resonance in the Electron Spin Resonance Line," *Physical Review*, vol. 103, pp. 834–835, 1956.

[15] L. J. Berliner, "The Evolution of Biomedical EPR (ESR)," *Biomedical Spectroscopy and Imaging*, vol. 5, pp. 5–26, 2016.

[16] D. T. Petasis and M. P. Hendrich, "A New Q-band EPR Probe for Quantitative Studies of Even Electron Metalloproteins," *Journal of Magnetic Resonance*, vol. 136, pp. 200–206, 1999.

[17] A. K. Upadhyay, D. T. Petasis, D. M. Arciero, A. B. Hooper and M. P. Hendrich, "Spectroscopic Characterization and Assignment of Reduction Potentials in the Tetraheme Cytochrome c554 from Nitrosomonas europaea," *Journal of the American Chemical Society*, vol. 125, pp. 1738–1747, 2003.

[18] R. Eisberg and R. Resnick, Quantum Physics of Atoms, Molecules, Solids, Nuclei, and Particles, New York: John Wiley & Sons, 1974.

[19] A. Morrish, The Physical Principles of Magnetism, New York: Wiley, 1965.

[20] J. W. Orton, Electron Paramagnetic Resonance: An Introduction to Transition Group Ions in Crystals, New York: Gordon and Breach Science Publishers Inc., 1969.

[21] G. Lancaster, "Electron Paramagnetic Resonance (A Review)," *Journal of Materials Science*, vol. 2, pp. 489–495, 1967.

2 Quantum mechanical background and formalism

Electron paramagnetic resonance (EPR) deals with the interaction of microwave radiation with a spin system of discrete energy states in the presence of an external magnetic field. The presence of discrete energy states in an atom or ion is consequently an essential component of the EPR phenomenon. The development of the theories that led to the existence of such states is a triumph of quantum mechanics, which provided an understanding of the physical properties of atoms and molecules. Even though some aspects of EPR can be explained using classical mechanics, the majority of EPR can only be understood with the use of quantum mechanics. This chapter is not a thorough discussion of quantum mechanics but rather a brief overview of select quantum mechanical principles that are essential for a proper understanding of EPR theory. More in-depth discussions of the quantum mechanics topics presented here can be found in the textbooks referenced at the end of the chapter.

2.1 The Schrödinger equation

The Schrödinger equation is a second-order differential equation that can provide information on the motion and energy states of a particle. The solution of the Schrödinger equation gives us the *wave function* $\Psi(x, y, z, t)$ of the particle, which carries information about the wavelike behavior of the particle. $\Psi(x, y, z, t)$ is not an observable – it has no physical meaning. It is a measure of the likelihood of finding the particle at coordinates (x, y, z) at time t. The measurable quantity is the probability density $|\Psi(x, y, z, t)|^2 = |\Psi^*(x, y, z, t)\Psi(x, y, z, t)|$ which measures the probability of finding the particle at (x, y, z) at time t. $\Psi^*(x, y, z, t)$ is the complex conjugate of $\Psi(x, y, z, t)$.

One-dimensional time-dependent Schrödinger equation for a free particle of mass m inside a potential $V(x)$ (this is actually a potential energy that in quantum mechanics is referred to as simply the potential function – not to be confused with the electric potential):

$$-\frac{\hbar^2}{2m}\frac{\partial^2\Psi(x, t)}{\partial x^2} + V(x)\Psi(x, t) = i\hbar\frac{\partial\Psi(x, t)}{\partial t} \qquad (2.1)$$

where $\hbar = h/2\pi = 6.58 \times 10^{-16}$ eVs. The time-dependent wave function $\Psi(x, t)$ can be separated into spatial $\psi(x)$ and time functions $f(t)$:

$$\Psi(x, t) = \psi(x)f(t) \qquad (2.2)$$

Substituting into the Schrödinger equation above, carrying out the derivatives and separating variables leads to

https://doi.org/10.1515/9783110417562-002

$$\frac{1}{\psi(x)}\left[-\frac{\hbar^2}{2m}\frac{d^2\psi(x)}{dx^2}+V(x)\psi(x)\right]=\frac{i\hbar}{f(t)}\frac{df(t)}{dt} \tag{2.3}$$

The two sides of this equation depend on entirely independent variables. The only time this can happen is when the two sides are equal to a constant, which we call E. This constant is actually the energy of the free particle giving a solution for the time-dependent function of

$$f(t) = e^{-iEt/\hbar} \tag{2.4}$$

that leads to the total wave function

$$\Psi(x, t) = \psi(x)e^{-iEt/\hbar} \tag{2.5}$$

The time-independent wave function is a solution of the time-independent Schrödinger equation

$$-\frac{\hbar^2}{2m}\frac{d^2\psi(x)}{dx^2}+V(x)\psi(x)=E\psi(x) \tag{2.6}$$

Solution of the Schrödinger equation subject to specific boundary conditions will give us values for the eigenfunctions ψ and the energy eigenvalues E.

2.2 The hydrogen atom

The hydrogen atom represents one of the major applications of the Schrödinger equation in examining the behavior of particles in three dimensions. Essentially, we want to know two things: the probability of finding the electron at a certain distance r from the proton (which is considered to be at rest at the center of coordinates) and determining the allowed energy states of the electron.

2.2.1 The Schrödinger equation in the hydrogen atom

For our purpose, solving the time-independent Schrödinger equation is sufficient in answering these questions – the time dependence does not yield anything of particular importance. The reason for this lies in the fact that the wave function

$$\Psi(x, t) = \psi(x)e^{-iEt/\hbar} \tag{2.7}$$

describing the electron in the hydrogen atom represents a stationary state since the probability density

$$|\Psi(x,\ t)|^2 = \psi^*(x)e^{iEt/\hbar}\psi(x)e^{-iEt/\hbar} = |\psi(x)|^2 \tag{2.8}$$

is independent of time.

In three dimensions, in a potential that is independent of time, the time-independent Schrödinger equation is given by

$$-\frac{\hbar^2}{2m}\nabla^2\psi + V\psi = E\psi \tag{2.9}$$

where

$$\nabla^2 \equiv \frac{\partial^2}{\partial x^2} + \frac{\partial^2}{\partial y^2} + \frac{\partial^2}{\partial z^2} \tag{2.10}$$

is the Laplacian in Cartesian coordinates. The potential that an electron with charge $-e$ at distance r from the nucleus with charge $+e$ experiences in the hydrogen atom is a central potential and only depends on r:

$$V(r) = -\frac{e^2}{r} \tag{2.11}$$

making it impossible to separate variables in the Schrödinger equation in Cartesian coordinates. Consequently, it makes more sense to convert the Schrödinger equation into spherical coordinates using the transformation relations $x = r\sin\theta\cos\phi$, $y = r\sin\theta\sin\phi$, $z = r\cos\theta$ as illustrated in Fig. 2.1. The volume element $d^3r = dxdydz$ in spherical coordinates is given by $d^3r = r^2 \sin\theta\ drd\theta d\phi$.

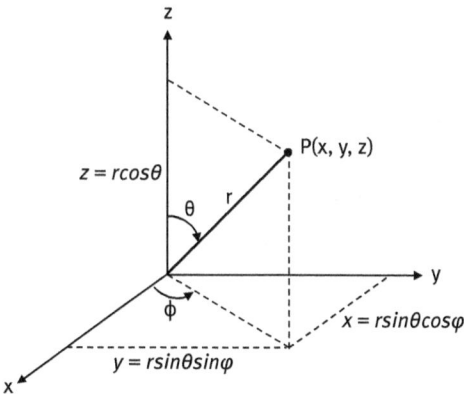

Fig. 2.1: Spherical coordinates (r, θ, ϕ) and their relationship to Cartesian coordinates (x, y, z).

The Schrödinger equation in spherical coordinates becomes

$$-\frac{\hbar^2}{2m}\left[\frac{1}{r^2}\frac{\partial}{\partial r}\left(r^2\frac{\partial}{\partial r}\right) + \frac{1}{r^2\sin\theta}\frac{\partial}{\partial\theta}\left(\sin\theta\frac{\partial}{\partial\theta}\right) + \frac{1}{r^2\sin^2\theta}\frac{\partial^2}{\partial\varphi^2}\right]\psi + V(r)\psi = E\psi \qquad (2.12)$$

where $\psi = \psi(r, \theta, \phi)$ [1]. We can use separation of variables to reduce this equation to three ordinary differential equations by expressing ψ as a product of functions that depend on the radial $R(r)$ and angular parts $Y(\theta, \phi)$:

$$\psi(r, \theta, \phi) = R(r)Y(\theta, \phi) \qquad (2.13)$$

leading to radial and angular equations that each has to be equal to a constant:

$$\frac{1}{R}\frac{d}{dr}\left(r^2\frac{dR}{dr}\right) - \frac{2mr^2}{\hbar^2}[V(r) - E] = \alpha \qquad (2.14)$$

$$-\frac{1}{Y}\left[\frac{1}{\sin\theta}\frac{\partial}{\partial\theta}\left(\sin\theta\frac{\partial Y}{\partial\theta}\right) + \frac{1}{\sin^2\theta}\frac{\partial^2 Y}{\partial\varphi^2}\right] = \alpha \qquad (2.15)$$

where the separation constant α is equal to $\alpha = l(l+1)$, where l is another constant to be defined later. Solutions for the radial equation are expressed in terms of specific constants labeled n and l and for the angular equation the constants are l and m_l. Consequently, the wave function is written as

$$\psi_{n,l,m_l}(r, \theta, \phi) = R_{n,l}(r)Y_l^{m_l}(\theta, \phi) \qquad (2.16)$$

to indicate the dependence on these constants. These constants are the quantum numbers for the hydrogen atom, and solutions exist only for specific integer values of these numbers, namely

$n = 1, 2, 3, \ldots$ (principal quantum number)
$l = 0, 1, 2, \ldots, n-1$ (azimuthal or angular momentum quantum number)
$m_l = 0, \pm1, \pm2, \ldots, \pm l$ (magnetic quantum number)

These numbers specify the energy and shape of the orbits of the electron around the nucleus. The solutions of the radial equation are given by

$$R(r) = r^l F(r)e^{-\sqrt{-2mE/\hbar^2}r} \qquad (2.17)$$

with physically acceptable solutions given when the function $F(r)$ is given in terms of the associated Laguerre polynomials. These are tabulated in many standard quantum mechanics textbooks [2]. The corresponding energy eigenvalues are given by

$$E_n = -\left[\frac{m}{2\hbar^2}\left(\frac{e^2}{4\pi\varepsilon_0}\right)^2\right]\frac{1}{n^2} = -\frac{13.6\ \text{eV}}{n^2} \qquad (2.18)$$

The radial probability density $r^2|R_{n,l}(r)|^2$ gives us the probability of where the electron can be found along the radial direction. These functions are plotted in Fig. 2.2 for $n = 1$, 2, 3 and $l = 0$, 1. The radial distance is expressed in terms of the Bohr radius a_0:

$$a_0 \equiv \frac{4\pi\varepsilon_0\hbar^2}{me^2} = 0.529\text{Å} \tag{2.19}$$

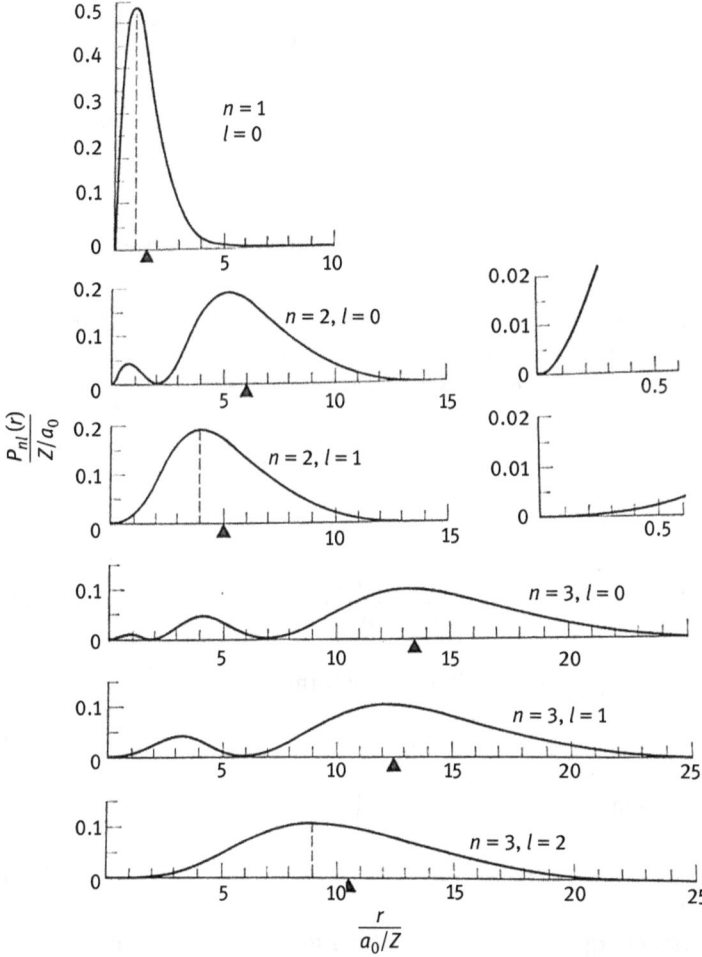

Fig. 2.2: Radial probability densities as a function of the distance r from the nucleus in terms of the Bohr radius a_0. The values of the quantum numbers n and l are shown for each plot. For each set of quantum numbers (n, l), the peak(s) indicate the region where the electron is most likely to be found along the radial direction. The triangles indicate the value of the expectation value of the radius $\langle r_{nl}\rangle$ [3].

The normalized wave functions $Y_l^{m_l}(\theta, \phi)$ of the angular equation are given by the spherical harmonics

$$Y_l^{m_l}(\theta, \phi) = \epsilon \sqrt{\frac{(2l+1)}{4\pi} \frac{(l-|m|)!}{(l+|m|)!}} \, e^{im_l\phi} P_l^{m_l}(\cos\theta) \tag{2.20}$$

where $\epsilon = (-1)^m$ for $m \geq 0$ and $\epsilon = 1$ for $m \leq 0$. $P_l^{m_l}(\cos\theta)$ are the associated Legendre polynomials. Plots of the probability density $|Y_l^{m_l}(\theta, \phi)|^2$ are shown in Fig. 2.3. These plots represent the angular probability distributions of the electron in three dimensions.

The complete wave function can be expressed as follows:

$$\psi_{nlm_l}(r, \theta, \phi) = \sqrt{\left(\frac{2}{na_0}\right)^3 \frac{(n-l-1)!}{2n[(n+l)!]^3}} \, e^{-r/na_0} \left(\frac{2r}{na_0}\right)^l [L_{n-l-1}^{2l+1}(2r/na_0)] Y_l^{m_l}(\theta, \phi) \tag{2.21}$$

where $L_{q-p}^p(x)$ are associated Laguerre polynomials tabulated in standard quantum mechanics textbooks such as reference [1].

In operator notation, the time-independent Schrödinger equation can be written in a compact way by using the *operator equivalents* of observables such as linear momentum whose components can be written as[1]

$$\hat{p}_x \rightarrow -i\hbar \frac{\partial}{\partial x}, \quad \hat{p}_y \rightarrow -i\hbar \frac{\partial}{\partial y}, \quad \hat{p}_z \rightarrow -i\hbar \frac{\partial}{\partial z} \tag{2.22}$$

We can now write the kinetic energy of a particle of mass m in operator form as

$$K = \frac{1}{2m}\left(p_x^2 + p_y^2 + p_z^2\right) = -\frac{\hbar^2}{2m}\left(\frac{\partial^2}{\partial x^2} + \frac{\partial^2}{\partial y^2} + \frac{\partial^2}{\partial z^2}\right) = -\frac{\hbar^2}{2m}\nabla^2 \tag{2.23}$$

which leads to the Schrödinger equation in operator form

$$\hat{\mathcal{H}}\psi = E\psi \tag{2.24}$$

where $\hat{\mathcal{H}}$ is the Hamiltonian operator

$$\hat{\mathcal{H}} = -\frac{\hbar^2}{2m}\nabla^2 + V \tag{2.25}$$

with V is the potential energy of the system, which for the hydrogen atom is the central potential $V(r)$ and the Laplacian ∇^2 is expressed in spherical coordinates. This form of the Schrödinger equation leads to the same expressions that we encountered earlier in this section.

---

[1] For the time-dependent SE we have to use the general wave function $\Psi(x, y, z, t)$ and the operator equivalent of energy $\hat{E} \rightarrow i\hbar \, \partial/\partial t$.

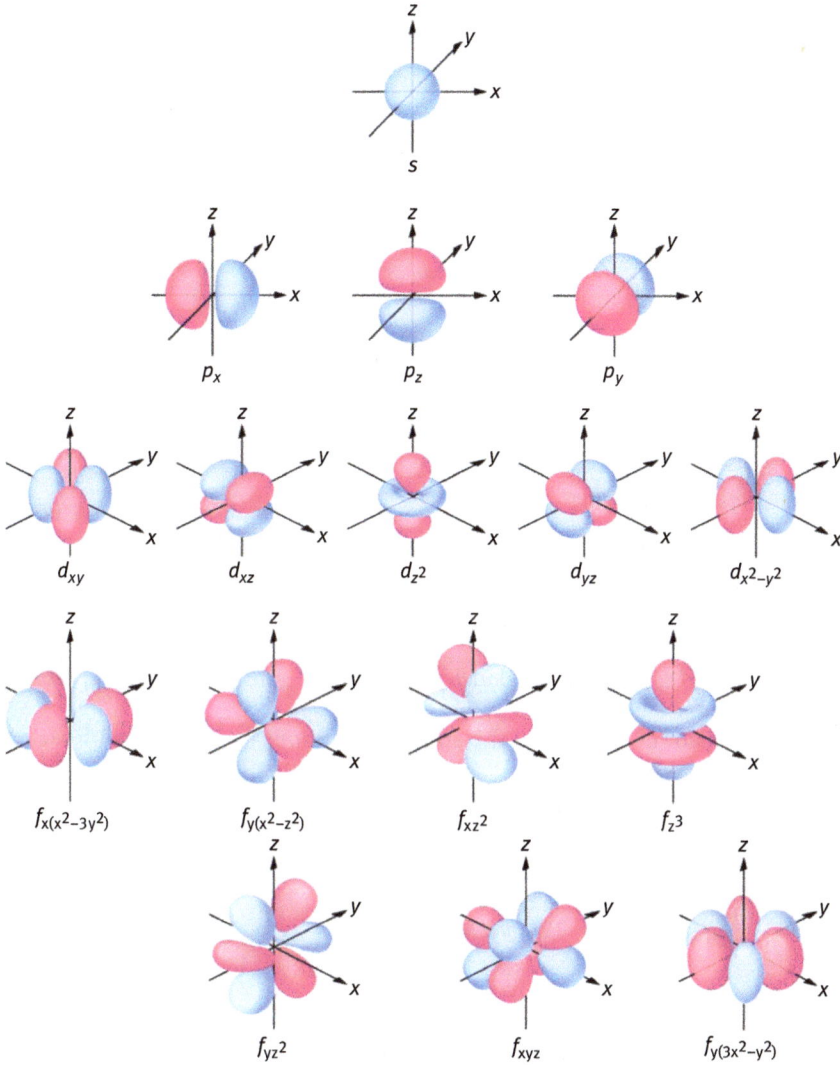

Fig. 2.3: Three-dimensional shapes of the s-, p-, d- and f-orbitals. These are diagrams of the probability density functions $\left[Y_l^{m_l}(\theta,\ \phi)\right]^*\left[Y_l^{m_l}(\theta,\ \phi)\right]$ whose shapes are determined by the associated Legendre polynomials. The three-dimensional shapes indicate the regions in space where there is large probability density. The different colors represent different phases of the wave function. They are also frequently represented by either a positive sign (blue) or a negative sign (pink) to indicate the sign of the wave function. Image reproduced from "Chemistry: Atoms First 2e" published by OpenStax and can be accessed for free at https://openstax.org/books/chemistry-atoms-first-2e/pages/1-introduction.

2.2.2 Orbital angular momentum

The classical expression for the orbital angular momentum

$$\vec{L} = \vec{r} \times \vec{p} \tag{2.26}$$

can be expressed in operator form by replacing the momentum components by their operator equivalents, eventually yielding the following angular momentum operator equivalents in spherical coordinates

$$\hat{L}_x = \frac{\hbar}{i}\left(-\sin\phi\frac{\partial}{\partial\theta} - \cos\phi\cot\theta\frac{\partial}{\partial\phi}\right) \tag{2.27}$$

$$\hat{L}_y = \frac{\hbar}{i}\left(\cos\phi\frac{\partial}{\partial\theta} - \sin\phi\cot\theta\frac{\partial}{\partial\phi}\right) \tag{2.28}$$

$$\hat{L}_z = \frac{\hbar}{i}\frac{\partial}{\partial\phi} \tag{2.29}$$

with

$$\hat{L}^2 = \hat{L}_x^2 + \hat{L}_y^2 + \hat{L}_z^2 = -\hbar^2\left[\frac{1}{\sin\theta}\frac{\partial}{\partial\theta}\left(\sin\theta\frac{\partial}{\partial\theta}\right) + \frac{1}{\sin^2\theta}\frac{\partial^2}{\partial\phi^2}\right] \tag{2.30}$$

The eigenfunctions of $\hat{L}^2$ and $\hat{L}_z$ are the spherical harmonics $Y_l^{m_l}(\theta, \phi)$ with eigenvalues $l(l+1)\hbar^2$ and $m_l\hbar$, respectively. This means that the three commuting operators $\hat{\mathcal{H}}$, $\hat{L}^2$, and $\hat{L}_z$ have simultaneous eigenfunctions

$$\hat{\mathcal{H}}\psi = E\psi \tag{2.31}$$

$$\hat{L}^2\psi = l(l+1)\hbar^2\psi \tag{2.32}$$

$$\hat{L}_z\psi = m_l\hbar\psi \tag{2.33}$$

Therefore, the orbital angular momentum is quantized and can only have certain orientations in space relative to an axis of quantization z, which implies that the orbit of the electron depends on the value of l. This result is called *space quantization* of $\vec{L}$ and is illustrated in Fig. 2.4 for the case of $l = 2$. The importance of this will become apparent in Chapter 4 when crystal field theory is discussed.

2.2.3 Orbital angular momentum in a magnetic field

The magnetic moment associated with the orbital angular momentum was already discussed in Chapter 1 and is given by

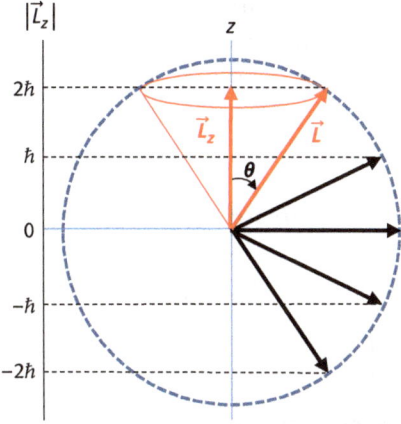

Fig. 2.4: Space quantization of $\vec{L}$: possible orbital angular momentum orientations in space for $l = 2$ relative to an axis of quantization defined by the z-axis. $\vec{L}$ has a constant magnitude equal to $|\vec{L}| = \sqrt{l(l+1)}\hbar = \sqrt{2(2+1)}\hbar = \sqrt{6}\hbar$ but its direction relative to the quantization axis can only have orientations that yield projections onto the z-axis given by $|\vec{L}_z| = L_z = m_l\hbar$. Since L_z is constant for a particular value of m_l, $\vec{L}$ can exist anywhere on the surface of an inverted cone with an axis of symmetry about z. The implication here is that we know the value of L_z exactly but have no knowledge about the components $\vec{L}_x$ and $\vec{L}_y$, which complies with the uncertainty principle.

$$\vec{\mu}_l = -\frac{e}{2m}\vec{L} = -g_l\frac{\beta}{\hbar}\vec{L} \tag{2.34}$$

where $g_l = 1$ is the orbital g-value. The negative sign indicates that $\vec{\mu}_l$ is in the opposite direction to that of $\vec{L}$. Classically, when a magnetic moment $\vec{\mu}$ is placed inside a magnetic field $\vec{B}$ it will experience a torque $\vec{\tau} = \vec{\mu} \times \vec{B}$ that will orient the magnetic moment parallel to the magnetic field to minimize the potential energy of interaction. In the case where there is an angular momentum associated with the magnetic moment, as in the case of the orbiting electron, the torque will be equal to

$$\vec{\tau} = \vec{\mu}_l \times \vec{B} = -\frac{g_l\beta}{\hbar}\vec{L} \times \vec{B} \tag{2.35}$$

This torque is perpendicular to the plane formed by $\vec{\mu}_l$ (or $\vec{L}$) and $\vec{B}$ as shown in Fig. 2.5, where an external magnetic field is applied at an angle θ relative to $\vec{L}$. This torque causes a change $d\vec{L}$ in $\vec{L}$ during a time interval dt since according to Newton's second law

$$\vec{\tau} = \frac{d\vec{L}}{dt} \tag{2.36}$$

Since the magnitude of $\vec{L}$ cannot change, then the direction of $\vec{L}$ must change. The result of this change $d\vec{L}$ is to cause $\vec{L}$ to precess about $\vec{B}$ with an angular velocity ω, an effect known as *Larmor precession*. According to Fig. 2.4 the projection $L_\perp$ of $\vec{L}$

onto the *xy*-plane will sweep through an angle ωdt in the time interval dt. We can see from the figure that the magnitude of $d\vec{L}$ is related to $L_\perp$ according to

$$dL = L_\perp \omega dt = L \sin \theta \omega dt \tag{2.37}$$

which yields

$$L \sin \theta \omega = \frac{dL}{dt} = \tau = \frac{g_l \beta}{\hbar} \left| \vec{L} \times \vec{B} \right| = \frac{g_l \beta}{\hbar} LB \sin \theta \tag{2.38}$$

giving an expression for the angular frequency of precession

$$\omega = \frac{g_l \beta}{\hbar} B \Rightarrow \vec{\omega} = \frac{g_l \beta}{\hbar} \vec{B} \tag{2.39}$$

which is called the *Larmor frequency*. It should be noted that a quantum mechanical calculation produces the same result for ω as the classically derived expression above. The same discussion can also be applied to the spin angular momentum inside a magnetic field by replacing $\vec{L}$ with $\vec{S}$ and $\vec{\mu}_l$ by $\vec{\mu}_s = -g_s(\beta/\hbar)\vec{S}$ as discussed in the previous chapter during the phenomenological discussion of EPR. A quantum mechanical discussion of the time evolution of the spin magnetic moment inside a magnetic field is given later in this section.

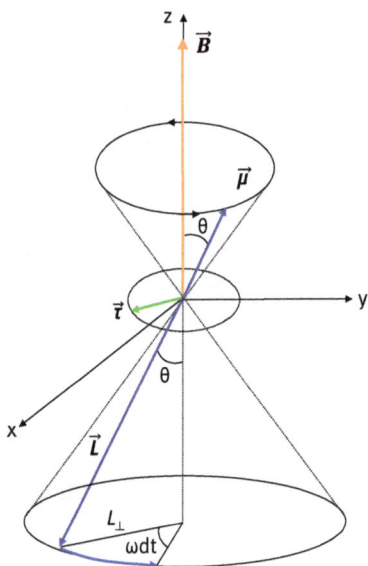

Fig. 2.5: Creation of a torque $\vec{\tau}$ when the magnetic moment of the electron $\vec{\mu}$ interacts with an external magnetic field $\vec{B}$. This torque gives rise to a precession of the orbital angular momentum $\vec{L}$ about $\vec{B}$ with frequency of precession equal to the Larmor frequency.

2.2.4 Spin angular momentum

The spin angular momentum $\vec{S}$ was initially introduced to explain experimental results (by Goudsmit and Uhlenbeck to explain the fine structure observed in line
spectra and the results of the Stern–Gerlach experiment) but was later introduced
by P. A. M. Dirac during the development of relativistic quantum mechanics that
culminated in the Dirac equation. The Dirac equation is a generalization of the
Schrödinger equation where the classical kinetic energy of the electron of mass m
in the hydrogen atom Hamiltonian given by

$$K = \boldsymbol{p}^2/2m \tag{2.40}$$

is replaced by the relativistic expression

$$K = \sqrt{(\boldsymbol{p}c)^2 + (mc^2)^2} - mc^2 \tag{2.41}$$

where mc^2 is the rest energy of the electron. Here we assume that the electron is
moving around an infinitely massive proton which is at rest. If we factor out the
term mc^2 we end up with

$$K = mc^2 \left(\sqrt{1 + \left(\frac{\boldsymbol{p}c}{mc}\right)^2} - 1 \right) \tag{2.42}$$

The hydrogen atom is a weakly relativistic system, since for an electron in the Bohr
model moving with speed v in a state with $n = 1$, the ratio

$$\frac{v}{c} = \frac{e^2}{\hbar c} = \frac{1}{137} \ll 1 \tag{2.43}$$

This result allows us to do a binomial expansion of the square root in eq. (2.42) to
get for the kinetic energy K the expression

$$K = mc^2 \left(1 + \frac{\boldsymbol{p}^2}{2m^2c^2} - \frac{\boldsymbol{p}^4}{8m^4c^4} + \cdots - 1 \right) = \frac{\boldsymbol{p}^2}{2m} - \frac{\boldsymbol{p}^4}{8m^3c^2} \tag{2.44}$$

keeping only the first three terms of the expansion. The first term is the nonrelativistic kinetic energy and the second term is the correction due to the variation of the
mass with the velocity in the relativistic regime [4]. The Hamiltonian for the hydrogen atom now becomes

$$\mathcal{H} = \frac{\boldsymbol{p}^2}{2m} + V(r) - \frac{\boldsymbol{p}^4}{8m^3c^2} + \cdots \tag{2.45}$$

where

$$V(r) = -\frac{q^2}{4\pi\varepsilon_0}\frac{1}{r} = -\frac{e^2}{r} \tag{2.46}$$

is the electrostatic potential energy between electron and proton ($e^2 = q^2/4\pi\varepsilon_0$) [4]. The first two terms in the Hamiltonian constitute the nonrelativistic Hamiltonian $\mathcal{H}_0$. The remaining term is the relativistic correction which is a perturbation on $\mathcal{H}_0$. The Hamiltonian $\mathcal{H}$ leads to the Dirac equation which fully incorporates relativistic effects with the addition of extra terms and, unlike the Schrödinger equation, leads to the theoretical prediction of the intrinsic spin. Full discussions of the Dirac equation can be found in more advanced quantum mechanics textbooks such as reference [5].

The magnetic moment associated with the spin angular momentum is given by

$$\vec{\mu}_s = -\frac{e}{m}\vec{S} = -g_s\frac{e}{2m}\vec{S} = -g_s\frac{\beta}{\hbar}\vec{S} \tag{2.47}$$

with the g-value equal to 2.0 as predicted by the Dirac equation (the observed value is actually equal to 2.00232, a discrepancy that is explained by quantum electrodynamics).

Quantum mechanically, we deal with the spin in terms of operators that act on the states $|s\, m_s\rangle$. The spin operators $\hat{S}_x$, $\hat{S}_y$ and $\hat{S}_z$ satisfy the commutation relations:

$$\left[\hat{S}_x, \hat{S}_y\right] = i\hbar\hat{S}_z \tag{2.48}$$

$$\left[\hat{S}_y, \hat{S}_z\right] = i\hbar\hat{S}_x \tag{2.49}$$

$$\left[\hat{S}_z, \hat{S}_x\right] = i\hbar\hat{S}_y \tag{2.50}$$

The operators $\hat{S}_z$ and

$$\hat{S}^2 = \hat{S}_x^{\,2} + \hat{S}_y^{\,2} + \hat{S}_z^{\,2} \tag{2.51}$$

commute:

$$\left[\hat{S}_z, \hat{S}^2\right] = 0 \tag{2.52}$$

consequently, the eigenvectors $|s\, m_s\rangle$ are simultaneous eigenstates of $\hat{S}_z$ and $\hat{S}^2$ with eigenvalues given by

$$\hat{S}^2|s\, m_s\rangle = s(s+1)\hbar^2|s\, m_s\rangle \tag{2.53}$$

$$\hat{S}_z|s\, m_s\rangle = m_s\hbar|s\, m_s\rangle \tag{2.54}$$

For an electron, the spin quantum number s is always equal to ½ with m_s only having two possible values equal to $\pm 1/2$. Consequently, there are two basis states for an electron given by $|\frac{1}{2}\ \frac{1}{2}\rangle$ (spin-up) and $|\frac{1}{2}\ -\frac{1}{2}\rangle$ (spin-down). These two states can be written in matrix form in terms of column vectors called *spinors*:

$$\psi_+ = \begin{pmatrix} 1 \\ 0 \end{pmatrix} \text{ spin-up} \tag{2.55}$$

$$\psi_- = \begin{pmatrix} 0 \\ 1 \end{pmatrix} \text{ spin-down} \tag{2.56}$$

A general state of an electron can then be expressed as follows:

$$\psi = \begin{pmatrix} a \\ b \end{pmatrix} = a\psi_+ + b\psi_- \tag{2.57}$$

The spin operators $\hat{S}_z$ and $\hat{S}^2$ can be expressed in terms of the Pauli spin matrices [4]

$$\sigma_x \equiv \begin{pmatrix} 0 & 1 \\ 1 & 0 \end{pmatrix}, \quad \sigma_y \equiv \begin{pmatrix} 0 & -i \\ i & 0 \end{pmatrix}, \quad \sigma_z \equiv \begin{pmatrix} 1 & 0 \\ 0 & -1 \end{pmatrix} \tag{2.58}$$

as

$$\hat{S}_x = \frac{\hbar}{2}\begin{pmatrix} 0 & 1 \\ 1 & 0 \end{pmatrix}, \quad \hat{S}_y = \frac{\hbar}{2}\begin{pmatrix} 0 & -i \\ i & 0 \end{pmatrix}, \quad \hat{S}_z = \frac{\hbar}{2}\begin{pmatrix} 1 & 0 \\ 0 & -1 \end{pmatrix} \tag{2.59}$$

$$\hat{S}^2 = \frac{3\hbar^2}{4}\begin{pmatrix} 1 & 0 \\ 0 & 1 \end{pmatrix} \tag{2.60}$$

Since the operators $\hat{S}_z$ and $\hat{S}^2$ commute, the eigenvectors ψ_+ and ψ_- are simultaneous eigenstates of $\hat{S}_z$ and $\hat{S}^2$ with eigenvalues given by

$$\hat{S}_z \chi_\pm = \pm \frac{\hbar}{2}\psi_\pm \tag{2.61}$$

$$\hat{S}^2 \chi_\pm = \frac{3\hbar^2}{4}\psi_\pm \tag{2.62}$$

A more convenient shorthand way of writing the spin states is given by the definitions:

$$\left|\frac{1}{2}\ \frac{1}{2}\right\rangle = \left|+\frac{1}{2}\right\rangle \quad \text{and} \quad \left|\frac{1}{2}\ -\frac{1}{2}\right\rangle = \left|-\frac{1}{2}\right\rangle \tag{2.63}$$

With this notation we can then write

$$\hat{S}_z \left| \pm \frac{1}{2} \right\rangle = \pm \frac{\hbar}{2} \left| \pm \frac{1}{2} \right\rangle \qquad (2.64)$$

$$\hat{S}^2 \left| \pm \frac{1}{2} \right\rangle = \frac{3\hbar^2}{4} \left| \pm \frac{1}{2} \right\rangle \qquad (2.65)$$

2.2.5 Electron spin inside a magnetic field

The magnetic dipole moment of the electron can be expressed in terms of the gyromagnetic ratio y as follows:

$$\vec{\mu} = y\vec{S} \qquad (2.66)$$

where

$$y = g\left(-\frac{e}{2m} \right) \qquad (2.67)$$

When placed inside a magnetic field $\vec{B}$ directed along the z-axis, the interaction energy can be expressed as a Hamiltonian

$$\mathcal{H} = -\vec{\mu} \circ \vec{B} = -yBS_z \qquad (2.68)$$

where S_z is the z-component of $\vec{S}$. Replacing S_z by its operator equivalent gives us the Hamiltonian operator

$$\hat{\mathcal{H}} = yB\hat{S}_z \qquad (2.69)$$

which allows us to determine the behavior and evolution of the electronic spin inside $\vec{B}$. First, we will discuss the possible energy states of the spin inside the magnetic field: The eigenvectors of $\hat{\mathcal{H}}$ are the same as those of $\hat{S}_z$ with eigenvalues given by

$$\hat{\mathcal{H}} \left| \pm \frac{1}{2} \right\rangle = \mp \frac{yB\hbar}{2} \left| \pm \frac{1}{2} \right\rangle \qquad (2.70)$$

This means that the spin-up and spin-down states of the electron spin have different energies given by

$$E_+ = -\frac{yB\hbar}{2} \qquad (2.71)$$

for the spin-up state and

$$E_- = +\frac{yB\hbar}{2} \qquad (2.72)$$

for the spin-down state as shown in Fig. 2.6. In other words, placing an electron inside an external magnetic field removes the twofold degeneracy of its eigenstates. The energy separation of these two states is

$$\Delta E = \gamma B \hbar \tag{2.73}$$

Using the Bohr magneton

$$\beta = \frac{e\hbar}{2m} \tag{2.74}$$

we can write

$$\frac{\gamma}{2} = \frac{g}{2}\left(-\frac{e}{2m}\right) = -\frac{1}{2}\frac{g\beta}{\hbar} \tag{2.75}$$

resulting in

$$E_{\pm} = \pm\frac{1}{2}g\beta B \tag{2.76}$$

Using $\Delta E = g\beta B$ gives us the *EPR resonance condition*

$$g\beta B = h\nu \tag{2.77}$$

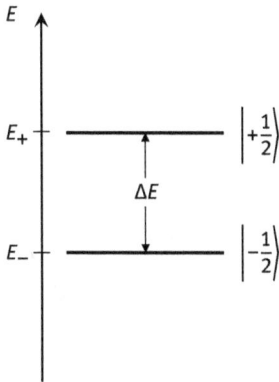

Fig. 2.6: Energy-level diagram for an electron spin inside a magnetic field directed along the z-axis. The energy separation between the two spin states is $\Delta E = g\beta B$.

The time-dependent Schrödinger equation can actually be used to discuss the time evolution of the spin [6]:

$$\hat{\mathcal{H}}|\psi\rangle = i\hbar\frac{\partial}{\partial t}|\psi\rangle \tag{2.78}$$

With the electron at rest, the total energy is given by

$$E = U = \vec{\mu}\cdot\vec{B} = \frac{2\beta}{\hbar}\vec{B}\cdot\vec{S} = \frac{2\beta}{\hbar}BS_z = \frac{2\beta}{\hbar}B\frac{\hbar}{2}\sigma_z = \beta B\sigma_z \tag{2.79}$$

where the magnetic field was taken along the z-axis. The Schrödinger equation then becomes

$$\beta B \sigma_z |\psi\rangle = i\hbar \frac{\partial}{\partial t} |\psi\rangle \tag{2.80}$$

Due to the definition of σ_z as given in the previous section, we already saw that we can write $|\psi\rangle$ in matrix form as

$$|\psi\rangle = \begin{pmatrix} \psi_+ \\ \psi_- \end{pmatrix} \tag{2.81}$$

giving rise to

$$\beta B \begin{pmatrix} 1 & 0 \\ 0 & -1 \end{pmatrix} \begin{pmatrix} \psi_+ \\ \psi_- \end{pmatrix} = i\hbar \frac{\partial}{\partial t} \begin{pmatrix} \psi_+ \\ \psi_- \end{pmatrix} \tag{2.82}$$

yielding the differential equations

$$\beta B \psi_+ = i\hbar \frac{\partial \psi_+}{\partial t} \tag{2.83}$$

$$-\beta B \psi_- = i\hbar \frac{\partial \psi_-}{\partial t} \tag{2.84}$$

with solutions

$$\psi_+ = a e^{-i\frac{\beta B}{\hbar} t} = a e^{\frac{i\gamma B t}{2}} \tag{2.85}$$

$$\psi_- = b e^{i\frac{\beta B}{\hbar} t} = b e^{-\frac{i\gamma B t}{2}} \tag{2.86}$$

where the constants a and b are determined from the initial conditions, and due to the normalization condition $\langle \psi|\psi \rangle = 1$, they are related by $|a|^2 + |b|^2 = 1$. We can then express a and b in terms of an angle θ such that $a = \cos(\theta/2)$ and $b = \sin(\theta/2)$ yielding the state

$$|\psi\rangle = \begin{pmatrix} \cos\frac{\theta}{2} e^{\frac{i\gamma B t}{2}} \\ \sin\frac{\theta}{2} e^{-\frac{i\gamma B t}{2}} \end{pmatrix} \tag{2.87}$$

Using this wave function to calculate the expectation value $\langle \vec{S} \rangle$ of $\vec{S}$:

$$\langle S_x \rangle = \langle \psi(t)|S_x|\psi(t)\rangle = \frac{\hbar}{2} \sin\theta \cos(\gamma B t) \tag{2.88}$$

$$\langle S_y \rangle = \langle \psi(t)|S_y|\psi(t)\rangle = \frac{\hbar}{2} \sin\theta \sin(\gamma B t) \tag{2.89}$$

$$\langle S_z \rangle = \langle \psi(t)|S_z|\psi(t)\rangle = \frac{\hbar}{2}\cos\theta \qquad (2.90)$$

This result indicates that $\langle \vec{S} \rangle$ makes an angle θ with respect to the z-axis and precesses about z with the Larmor frequency $\omega = \gamma B$, which is the same result we derived for the precession of $\vec{L}$ using classical arguments.

2.3 Perturbation theory

Before discussing spin–orbit coupling and the Zeeman effect, we need to introduce perturbation theory, since, depending on our system, one of these effects can be much stronger than the other necessitating the treatment of the other effect as a perturbation on the first. Perturbation theory is also important in deciding the strength of the various terms that appear in the spin Hamiltonian.

We will discuss the time-independent nondegenerate perturbation theory, where the exact solutions of the Schrödinger equation are known for a particular unperturbed system. We will outline a theory below that allows us to get approximate solutions when a small perturbation is added to the system.[2] Suppose that the wave functions for the unperturbed system are given by ψ_n^0, and they are solutions of the Schrödinger equation

$$\mathcal{H}^0 \psi_n^0 = E_n^0 \psi_n^0 \qquad (2.91)$$

where $\mathcal{H}^0$ is the unperturbed Hamiltonian of the system. The eigenfunctions ψ_n^0 are orthonormal

$$\langle \psi_n^0|\psi_m^0 \rangle = \delta_{nm} \qquad (2.92)$$

with eigenvalues E_n^0 (δ_{nm} is the Kronecker delta[3]). We now consider a new problem where the Hamiltonian is given by

$$H = \mathcal{H}^0 + \mathcal{H}' \qquad (2.93)$$

where $\mathcal{H}^0$ is the unperturbed Hamiltonian and $\mathcal{H}'$ is the small perturbation acting on the system which is very small compared to $\mathcal{H}^0$: $\mathcal{H}' \ll \mathcal{H}^0$. The equation we need to solve now is

$$H\psi_n = E_n\psi_n \qquad (2.94)$$

---

2 The discussion in this section is an outline of a more extensive discussion given in Chapter 6 of reference [1].

3 The Kronecker delta $\delta_{nm} = 1$ when $n = m$ and 0 when $n \neq m$.

The new eigenfunctions ψ_n and eigenvalues E_n can be expressed as

$$\psi_n = \psi_n^0 + \psi_n^1 + \psi_n^2 + \cdots \tag{2.95}$$

and

$$E_n = E_n^0 + E_n^1 + E_n^2 + \cdots \tag{2.96}$$

where ψ_n^q and E_n^q are the qth-order corrections to the nth eigenfunction and the nth eigenvalue, respectively ($q > 0$). The matrix elements of $\mathcal{H}$ then follow from

$$\int \psi_m^{0*}\, \mathcal{H}\psi_n^0 d\tau = E_n^0 \delta_{nm} + H_{mn} \tag{2.97}$$

resulting in a matrix with diagonal elements H_{mn} from $\mathcal{H}^0$ and with both diagonal and off-diagonal elements from $\mathcal{H}^1$. To a first approximation, we neglect the off-diagonal elements so that the eigenvalues of $\mathcal{H}$ are given by the solutions of the block-diagonal terms. The first-order correction to the energy eigenvalues can be derived in terms of the unperturbed eigenfunctions as

$$E_n^1 = \langle \psi_n^0 | \mathcal{H}' | \psi_n^0 \rangle \tag{2.98}$$

As an illustration, Orton [7] consider the unperturbed state to comprise a threefold degenerate level with energies E_1^0, a twofold degenerate level with energies E_2^0 and another twofold degenerate level with energies E_3^0. The resulting block-diagonal matrix, where we ignored the off-diagonal terms, looks like

$$\begin{pmatrix} E_1^0 + H_{11} & H_{12} & H_{13} & & & & \\ H_{21} & E_1^0 + H_{22} & H_{23} & & & & \\ H_{31} & H_{32} & E_1^0 + H_{33} & & & & \\ & & & E_2^0 + H_{44} & H_{45} & & \\ & & & H_{54} & E_2^0 + H_{55} & & \\ & & & & & E_3^0 + H_{66} & H_{67} \\ & & & & & H_{76} & E_3^0 + H_{77} \end{pmatrix} \tag{2.99}$$

Diagonalizing the block-diagonal terms yields the first-order correction to the energy E_n^1.

The perturbed eigenfunctions ψ_n^1 can be expressed as a linear combination of the unperturbed wave functions, with the exception of ψ_n^0, according to[4]

---

4 The unperturbed wave functions form a complete set. According to Dirichlet's theorem, any function $f(x)$ can be written as a linear combination of functions $\psi_i(x)$ that form a complete set: $f(x) = \sum_{i=1}^{\infty} c_i\, \psi_i(x)$.

$$\psi_n^1 = \sum_{m \neq n} c_{nm} \psi_m^0 \tag{2.100}$$

The coefficients c_{nm} can be calculated from the perturbed Hamiltonian:

$$c_{nm} = \frac{\langle \psi_m^0 | \mathcal{H}' | \psi_n^0 \rangle}{(E_n^0 - E_m^0)} \tag{2.101}$$

which yields the first-order correction to the perturbed eigenfunctions:

$$\psi_n^1 = \sum_{m \neq n} \frac{\langle \psi_m^0 | \mathcal{H}' | \psi_n^0 \rangle}{(E_n^0 - E_m^0)} \psi_m^0 \tag{2.102}$$

This function is fine as long as the unperturbed system has no degenerate energy states. If there is degeneracy, then there are issues with the denominator where we could be dividing by zero. Such issues are avoided in *degenerate perturbation theory*, which will not be discussed here since our applications of perturbation theory will only involve singlet states.

Writing the perturbed Hamiltonian to include corrections up to second order in ψ and E allows the derivation of the second-order correction to the energy given by the equation

$$E_n^2 = \sum_{m \neq n} \frac{\langle \psi_m^0 | \mathcal{H}' | \psi_n^0 \rangle \langle \psi_n^0 | \mathcal{H}' | \psi_m^0 \rangle}{(E_n^0 - E_m^0)} = \sum_{m \neq n} \frac{|\langle \psi_m^0 | \mathcal{H}' | \psi_n^0 \rangle|^2}{(E_n^0 - E_m^0)} \tag{2.103}$$

The last two expressions indicate that the perturbation causes a *mixing* of the state ψ_n^0 with the other eigenstates of $\mathcal{H}^0$. If the perturbation $\mathcal{H}'$ has no matrix element between a state ψ_m^0 and ψ_n^0, then the contribution of that state to the first-order correction to the wave function and second-order correction to the energy is zero. An application of this will be seen in the discussion of spin–orbit coupling in the next section and in Chapter 3.

This short outline of perturbation theory will end here since no higher order terms will be needed in the material discussed in this book.

2.4 Spin–orbit coupling

Spin–orbit coupling arises from the interaction of an electron with the magnetic field produced by the positive nucleus as perceived in the frame of reference of the electron (Fig. 2.7(a) and (b)). In the Bohr model of the hydrogen atom, the orbiting electron of mass m and charge $-e$ has an orbital angular momentum $\vec{L}$ due to its circular motion of radius r around the proton. The electron also has a spin angular momentum $\vec{S}$ and an associated magnetic moment $\vec{\mu}_S$ which can have two possible orientations relative to $\vec{L}$. The interaction of the magnetic field $\vec{B}$ produced by the motion of the nucleus about the electron and the magnetic moment $\vec{\mu}_S$ results in two possible energies for the electron as shown in Fig. 2.7(c). This effect, along with the relativistic correction to the hydrogen atom Hamiltonian $\mathcal{H}_0$, contributes to the origin of the *fine structure* observed in the optical spectra of the hydrogen atom.

(a)

(b)

(c)

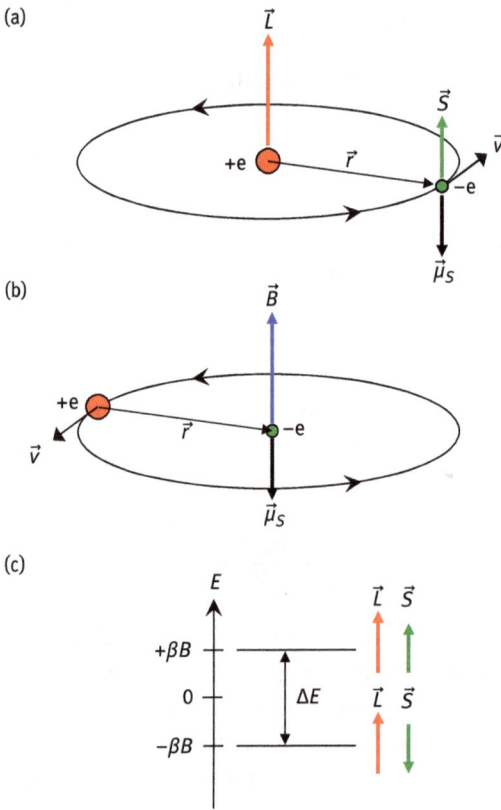

Fig. 2.7: Origin of spin–orbit coupling in the hydrogen atom. (a) The electron with linear velocity $\vec{v}$ and spin angular momentum $\vec{S}$ orbiting the stationary nucleus creating an orbital angular momentum $\vec{L}$. (b) In the electron frame of reference, the nucleus orbits the electron producing a magnetic field $\vec{B}$ at the electron. (c) This magnetic field interacts with the magnetic moment of the electron giving rise to two possible energies of the electron due to the two possible orientations of $\vec{S}$ relative to $\vec{L}$.

As shown in Fig. 2.8(a), the angular momentum of the electron with velocity $\vec{v}_e$ is equal to

$$\vec{L} = \vec{r} \times (m\vec{v}_e) \tag{2.104}$$

with direction along the positive z-axis. In the reference frame of the electron (Fig. 2.8(b)), the proton (charge $q = +e$) orbits the electron in a circular motion with velocity $\vec{v}_n = -\vec{v}_e$ producing a magnetic field at the electron equal to

$$\vec{B} = \frac{\mu_0}{4\pi} \frac{q\vec{v}_n \times \vec{r}}{r^3} = \frac{1}{c^2} \frac{e\vec{v}_n \times \vec{r}}{r^3} = -\frac{e}{c^2} \frac{\vec{v}_e \times \vec{r}}{r^3} = \frac{e}{c^2} \frac{\vec{r} \times \vec{v}_e}{r^3} \tag{2.105}$$

where the relationship $c^2 = 1/\varepsilon_0\mu_0$ (c is the speed of light) was used, and the multiplicative constant $1/4\pi\varepsilon_0$ was set equal to 1. The direction of $\vec{B}$ is along the positive z-axis. Replacing the cross product by $\vec{L}/m$ results in the magnetic field

$$\vec{B} = \frac{e}{mc^2r^3}\vec{L} \tag{2.106}$$

This magnetic field interacts with the spin magnetic moment of the electron $\vec{\mu} = -(e/m)\vec{S}$ with a potential energy of interaction equal to

$$E = -\vec{\mu} \cdot \vec{B} = \left(\frac{e^2}{m^2c^2r^3}\right)\vec{L} \cdot \vec{S} \tag{2.107}$$

This expression is actually off by a factor of 1/2, which is correctly written as

$$E = \frac{1}{2}\left(\frac{e^2}{m^2c^2r^3}\right)\vec{L} \cdot \vec{S} \tag{2.108}$$

The factor of 1/2 arises from the fact that the electron is accelerating as it moves in a circular orbit around the nucleus (the electron frame of reference is a noninertial frame of reference) and it is explained by a relativistic effect called *Thomas precession* [3].

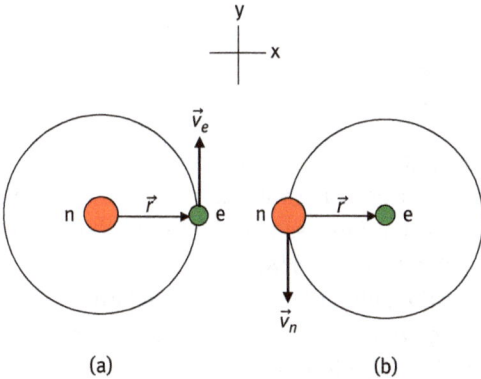

(a) (b)

Fig. 2.8: (a) An electron (e) moving in a circular Bohr orbit of radius $\vec{r}$ and velocity $\vec{v}_e$ around the nucleus (n) as observed in the frame of reference of the nucleus. (b) The same motion as observed in the frame of reference of the electron. The nucleus moves with a velocity $\vec{v}_n = -\vec{v}_e$ about the electron creating a magnetic field $\vec{B}$ at the location of the electron along the positive z-axis.

This expression gives rise to the spin–orbit coupling Hamiltonian for the electron in the hydrogen atom

$$\mathcal{H}_{SO} = \zeta \vec{L} \cdot \vec{S} \tag{2.109}$$

where the parameter ζ given by

$$\zeta = \frac{e^2}{2mc^2 r^3} \tag{2.110}$$

is a measure of the strength of the interaction between the spin of a single electron and its orbital angular momentum.

For multielectron atoms, the above expression is multiplied by the atomic number Z, indicating that spin–orbit coupling is more important for heavier elements. We should keep in mind that the expression for ζ is an approximation that does not take into account the true nature of the diffuse orbit of the electron and the shielding of the nuclear charge by the electronic orbitals in multielectron atoms. A more accurate expression for ζ is given by

$$\zeta = \frac{Z_{eff}e^2}{2mc^2 \langle r^3 \rangle} \tag{2.111}$$

where $\langle r^3 \rangle$ is the average of r^3 and $Z_{eff}e$ is the effective nuclear charge.

The spin–orbit coupling Hamiltonian for multielectron atoms is generally expressed in terms of a constant λ which is a property of the energy level instead of a single electron:

$$\mathcal{H}_{SO} = \lambda \vec{L} \cdot \vec{S} \tag{2.112}$$

where

$$\lambda = \pm \frac{\zeta}{2S} \tag{2.113}$$

with λ called the spin–orbit coupling constant.[5] ζ is a positive quantity, as we have seen earlier, but λ can have both positive and negative values. In transition series ions, the positive sign for λ applies to shells that are less than half-full (d^1–d^4) and the negative sign for more than half-full orbitals (d^6–d^{10}) [8].

Spin–orbit coupling is responsible for the shifts of the g-values relative to the free-electron value g_e. The deviations of g from g_e are large for ions with large spin–orbit coupling constants ζ. In some of the lighter transition series ions, this interaction is quite small (due to the quenching or partial quenching of $\vec{L}$ by the crystal

---

5 In EPR literature, the constants λ and ζ are both referred to as the spin–orbit coupling constant by different authors.

field which will be explained later) but it can be substantial in heavier transition series. ζ has large values for rare earth ions as well due to the fact that the magnetic electrons are inner 4f electrons. Values of ζ for some free transition ions with a single electron in a d-orbital are tabulated in Tab. 2.1 to illustrate these differences.

Tab. 2.1: Spin–orbit coupling constants ζ for select single-electron transition ions [9].

Z^+ / M	0	1^+	2^+	3^+	4^+	5^+	6^+
Cr	135	185	230	275	355	380	
Mn	190	255	300	355	415	475	540
Fe	275	335	400	460	520	590	665
Co	390	455	515	580	650	715	790
Ni		565	630	705	790	865	950
Cu			830	890	960	1,030	1,130

The spin–orbit interaction is added as another relativistic correction term in the hydrogen atom Hamiltonian of eq. (2.45): Both correction terms can be treated with perturbation theory to determine the effect they have on the energy levels of the hydrogen atom:

$$\mathcal{H} = \frac{p^2}{2m} + V(r) - \frac{p^4}{8m^3c^2} + \frac{e^2}{2mc^2r^3}\vec{L}\cdot\vec{S}\cdots \tag{2.114}$$

To compare the spin–orbit term to the unperturbed nonrelativistic Hamiltonian $\mathcal{H}_0$: $\vec{L}$ and $\vec{S}$ are each of order $\hbar$ and $\mathcal{H}_0$ is of order e^2/r yielding [4]

$$\frac{\mathcal{H}_{SO}}{\mathcal{H}_0} \sim \frac{\frac{e^2}{mc^2r^3}}{\frac{e^2}{r}} = \frac{e^4}{\hbar^2c^2} = \left(\frac{1}{137}\right)^2 \tag{2.115}$$

where r was taken to be of the order of the Bohr radius $a_0 = \hbar^2/me^2$. The treatment of the spin–orbit term as a perturbation of $\mathcal{H}_0$ is, therefore, justified. This will allow us later on to treat spin–orbit coupling as a perturbation to other effects such as electron–electron and crystal-field interactions.

The effect of the first relativistic correction term (also referred to as the *mass–velocity* term) on the hydrogen energy levels E_n can be obtained by treating it as a perturbing relativistic Hamiltonian

$$\mathcal{H}'_r = -\frac{p^4}{8m^3c^2} \tag{2.116}$$

and using first-order perturbation theory to get the energies[6]

$$E_r^1 = \left\langle \mathcal{H}_r' \right\rangle = \left\langle \psi_n^0 | \mathcal{H}_r' | \psi_n^0 \right\rangle = -\frac{E_n^2}{2mc^2} \left(\frac{4n}{l+\frac{1}{2}} - 3 \right) \tag{2.117}$$

which indicates that there is a very small shift in the unperturbed energy levels E_n.

2.5 Total angular momentum

Because of the existence of spin–orbit coupling, $\vec{L}$ and $\vec{S}$ no longer commute with the total Hamiltonian of the system and are not constants of the motion. Their vector sum

$$\vec{J} = \vec{L} + \vec{S} \tag{2.118}$$

however does commute with the total Hamiltonian of the system and is a constant of the motion. $\vec{J}$ is defined as the total angular momentum of the electron. Classically, a coupling term of the form $\vec{L} \cdot \vec{S}$ can be expressed as

$$\mathcal{H}_{SO} = \lambda \vec{L} \cdot \vec{S} = \lambda LS \cos \theta \tag{2.119}$$

where θ is the angle between $\vec{L}$ and $\vec{S}$. When $\mathcal{H}_{SO}$ is much smaller than the Hamiltonian $\mathcal{H}_0$, the result is precession of $\vec{L}$ and $\vec{S}$ about $\vec{J}$ with an angular velocity proportional to $\lambda |\vec{J}|$ [4]. In the presence of an external magnetic field $\vec{B}$ along the z-axis, $\vec{J}$ will precess about the z-axis. The magnitude and z-component of $\vec{J}$ are expressed in terms of the quantum numbers j and m_j:

$$J = \sqrt{j(j+1)}\hbar \tag{2.120}$$

and

$$J_z = m_j \hbar \tag{2.121}$$

with $m_j = -j, -j+1, \ldots, 0, \ldots, j-1, j$ and $j = l+s, l+s-1, \ldots, |l-s| = l+1/2, l-1/2$ when $s = 1/2$.

We can now determine the effect of spin–orbit coupling on the hydrogen atom energy levels by writing $\vec{L} \cdot \vec{S}$ in eq. (2.112) as

$$\vec{L} \cdot \vec{S} = \frac{1}{2} \left(J^2 - L^2 - S^2 \right) \tag{2.122}$$

---

6 Complete calculation of this can be found in Chapter 6 of reference [1].

since the Hamiltonian commutes with J, L^2, S^2. The eigenvalues of $\vec{L} \cdot \vec{S}$ with $s = 1/2$ are

$$\frac{\hbar^2}{2} [j(j+1) - l(l+1) - s(s+1)] \tag{2.123}$$

Treating the spin–orbit term as a perturbing Hamiltonian

$$\mathcal{H}'_{so} = \frac{e^2}{2mc^2 r^3} \vec{L} \cdot \vec{S} \tag{2.124}$$

first-order perturbation theory allows us to get the correction to the hydrogen energy levels

$$E^1_{so} = \left\langle \mathcal{H}'_{so} \right\rangle = \left\langle \psi^0_n | \mathcal{H}'_{so} | \psi^0_n \right\rangle = \frac{E_n^2}{mc^2} \left[n \frac{j(j+1) - l(l+1) - \frac{3}{4}}{l(l + \frac{1}{2})(l+1)} \right] \tag{2.125}$$

where $s = 1/2$ was used. This indicates that spin–orbit coupling also causes a small shift in the hydrogen energy levels E_n, which is of the same order of magnitude as the shifts produced by the relativistic correction term.[7] Adding this result to the expression for the relativistic correction E^1_r leads to the overall energy shift that is responsible for the fine structure observed in the emission spectra of the hydrogen atom[8]

$$E^1_{fs} = \frac{E_n^2}{2mc^2} \left(3 - \frac{4n}{j + \frac{1}{2}} \right) \tag{2.126}$$

The total energy of the hydrogen levels with principal quantum number n and total quantum number j is therefore given by

$$E_{nj} = -E_n \left[1 + \frac{\alpha^2}{n^2} \left(\frac{n}{j + \frac{1}{2}} - \frac{3}{4} \right) \right] \tag{2.127}$$

where E_n are the hydrogen energies from page 4 and $\alpha \approx 1/137$ is the fine-structure constant. Fine structure results in the lifting of the degeneracy of the l states in a particular n shell. The degeneracy in j, however, is not lifted. The net effect on the hydrogen energy levels is shown in Fig. 2.9.

---

7 It should be noted that spin–orbit coupling is also a relativistic effect, but from a different mechanism than the earlier relativistic correction.

8 The first relativistic correction term (the *mass–velocity* term) is more important for hydrogen and for atoms with small atomic number Z. We will see later that for multielectron atoms, the spin–orbit interaction becomes more important and dominates for atoms with larger Z. Consequently, for transition series atoms that form a large part of the systems studied with EPR we will essentially be speaking of spin-orbit coupling instead of fine-structure.

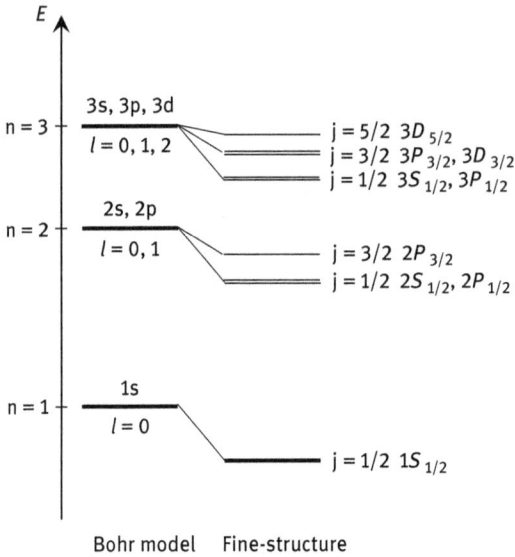

Fig. 2.9: Effect of fine structure on the energy levels of the hydrogen atom (not to scale). Fine structure lifts the degeneracy in l but not in j. The fine structure states are labeled according to the spectroscopic notation nL_j where $L = S$, P, D, ... corresponding to the values $l = 0$, 1, 2, ... The energy shifts are greatly exaggerated in this diagram.

2.6 Multielectron atoms

Compared to the hydrogen atom, multielectron atoms are very complicated. The relatively simple case of a single particle (electron) interacting only with one other particle (the nucleus) cannot be applied here since many-electron atoms involve many interacting particles. In an atom with atomic number Z, each of the Z electrons experiences the strong, attractive Coulomb potential from the nucleus with charge $+Ze$ but also experiences the repulsive potential from the remaining $(Z - 1)$ electrons each with charge $-e$. The presence of interactions between the electrons makes this a difficult problem. Some rather elaborate approximation techniques are employed to determine the behavior of the electronic levels, which are important in EPR. Whole books are written on the subject but here we will just give a very brief outline of the methods used to get the energy levels of multielectron atoms. We start by ignoring electronic interactions and placing the electrons on hydrogen-like states called *orbitals* subject to the Pauli exclusion principle, which states that no two electrons can exist on the same state with the exact same set of quantum numbers. This results in a maximum of two electrons occupying each orbital having opposite spins.[9] These orbitals have the same energy E_n which depends on the principal quantum number n as we have already seen. Such states are called *shells* with degeneracy $2(2l+1)$ and a maximum electron occupancy of $2n^2$.

When interactions between the electrons are taken into account, we no longer have a central potential. The nonrelativistic multielectron Hamiltonian with a stationary nucleus can be expressed as

$$\mathcal{H} = \sum_{i=1}^{Z}\left(-\frac{\hbar^2}{2m}\vec{\nabla}_i^{\,2} - \frac{Ze^2}{r_i}\right) + \sum_{i<j}\frac{e^2}{|\vec{r}_i - \vec{r}_j|} \tag{2.128}$$

The first summation represents the kinetic energy of individual electrons and the Coulomb attraction between each electron and the nucleus. The second summation represents the Coulomb repulsion between electrons i and j separated by a distance $|\vec{r}_i - \vec{r}_j|$ (electron–electron repulsion). The resulting Schrödinger equation cannot be solved exactly to give us the energies and wave functions of the atom, even for He, the simplest multielectron atom. Such *many-body* problems can only be solved using approximation methods. Even treating the electron interaction term with perturbation theory does not yield acceptable results since the separation distance $|\vec{r}_i - \vec{r}_j|$ between the electrons is of the same order of magnitude as the distance between electrons and the nucleus. We can, however, make use of a *central-field approximation* when we consider that an inner electron will experience a stronger Coulomb field from the nucleus compared to an outer electron where shielding of the nuclear charge occurs by the diffuse cloud of the remaining $(Z-1)$ electrons. An inner electron will essentially experience the full nuclear potential, whereas an outer electron will experience a nuclear potential of an effective charge $+e$, and electrons between these two limits will experience a potential that does not have a $1/r_i$ dependence:

$$V_c(r_i) = -\frac{Ze^2}{r_i} \text{ for small } r_i \tag{2.129}$$

$$V_c(r_i) = -\frac{e^2}{r_i} \text{ for large } r_i \tag{2.130}$$

This potential allows us to express the Hamiltonian in a form that can be manipulated using the Hartree variational method where the wave function can be written as a product of hydrogen wave functions

---

9 Electrons are *fermions*: indistinguishable particles of half-integer spin that obey Fermi–Dirac statistics. The Pauli exclusion principle is a result of the two-particle wave function of two electrons on the same energy state being antisymmetric.

$$\psi(\vec{r}_1, \vec{r}_2, \vec{r}_3, \dots) = \psi_{n_1 l_1 m_{l_1}}(\vec{r}_1)\psi_{n_2 l_2 m_{l_2}}(\vec{r}_1)\psi_{n_3 l_3 m_{l_3}}(\vec{r}_1)\cdots \qquad (2.131)$$

and substituted as trial function into the Schrödinger equation. The exclusion principle is satisfied here since each wave function is given different sets of quantum numbers. Applying the variational method allows us to get an estimate on the energies of a multielectron atom.[10] A more accurate method called the Hartree–Fock method applies the variational method to a wave function that is expressed in terms of a Slater determinant and results in better energy estimates. It is sufficient for our purposes to state here that the overall result is a removal of the degeneracy of the l states within shells: within a particular shell, the energy increases with increasing l. The arrangement of the energy levels in multielectron atoms is shown in Fig. 2.10.

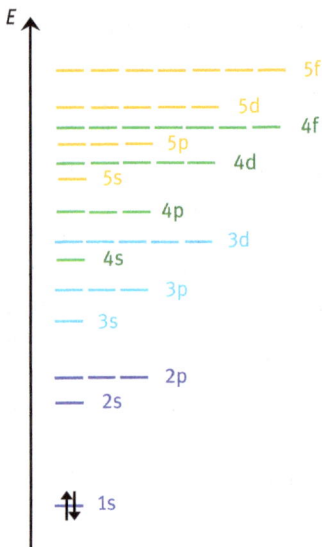

Fig. 2.10: Arrangement of energy levels in a multielectron atom according to the Hartree theory. Each line represents a doubly degenerate state, which can accommodate two electrons that obey the Pauli exclusion principle. For each level, the total degeneracy is equal to $2(2l+1)$. The figure is not drawn to scale – energy scale between certain levels is actually very small but is exaggerated here to better distinguish the states.

2.7 Total angular momentum in multielectron atoms

The interaction between the orbital and spin angular momenta in many-electron atoms leads to the definition of a total angular momentum that can be illustrated within the context of spin–orbit coupling.

---

10 An illustration of the use of the variational method is given in Chapter 7 of reference [1], where it is applied to the case of the He atom.

2.7.1 LS or Russell–Saunders coupling

In the case of weak spin–orbit coupling, the total angular momentum is determined through the Russell–Saunders (or LS) coupling scheme, which is valid for most transition series ions other than the very heavy ions. In this scheme, the Coulomb interactions between atoms on orbits with different l quantum numbers (orbit–orbit interactions) are stronger than the relativistic spin–orbit interactions that couple the l and s quantum numbers. In this case, the individual orbital and spin angular momenta can be added up independently from one another to give rise to resultant vectors $\vec{L}$ and $\vec{S}$ defined as

$$\vec{L} = \sum_i \vec{L_i} \ \text{ and } \ \vec{S} = \sum_i \vec{S_i} \tag{2.132}$$

where the summations are taken over the electrons on a particular shell. The individual momenta are added up vectorially keeping in mind the quantization rules. For example, in the interaction of two electrons, the summation $\vec{L_1} + \vec{L_2}$ yields the values

$$L = L_1 + L_2, \qquad L_1 + L - 1, \ \ldots, \ |L_1 - L_2| \tag{2.133}$$

with

$$m_L = -L, \ -L + 1, \ \ldots, \ 0, \ \ldots, \ L - 1, \ L \tag{2.134}$$

For the spin addition, $\vec{S_1} + \vec{S_2}$ yields

$$S = S_1 + S_2, \ S_1 - S_2 \tag{2.135}$$

with

$$m_S = -S, \ -S + 1, \ \ldots, \ 0, \ \ldots, \ S - 1, \ S \tag{2.136}$$

$\vec{L}$ and $\vec{S}$ are weakly coupled via spin–orbit coupling to give a total angular momentum

$$\vec{J} = \vec{L} + \vec{S} \tag{2.137}$$

with the total angular momentum quantum number given by

$$J = L + S, \ L + S - 1, \ \ldots, \ |L - S| \tag{2.138}$$

with

$$|\vec{J}| = \sqrt{J(J+1)}\,\hbar \tag{2.139}$$

and

$$J_z = m_j \hbar \tag{2.140}$$

where

$$m_J = -J, -J+1, \ldots, 0, \ldots, J-1, J \qquad (2.141)$$

This example is illustrated in Fig. 2.11.

2.7.2 Hund's rules

Hund's rules are a set of conditions that allow the determination of the ground state configuration of electrons in atomic shells and work best for Russell–Saunders coupling. There are three rules for the electron occupied states [10]:

(1) Obtain the maximum total atomic spin without violating the Pauli exclusion principle.

(2) Maximize the total angular momentum while remaining consistent with (1).

(3) The total angular momentum $\vec{J}$ has the value

$J = |L - S|$ when shell is less than half-full

$J = L + S$ when shell is more than half-full

$J = S$ for half-full shell

The ground state is then designated by the term $^{2S+1}L_J$ where L is represented by the spectroscopic symbols S, P, D, F for the values $L = 0, 1, 2, 3$, respectively.

Example: Cr^{3+} is a d^3 ion. Following the first rule, the electrons are distributed on the available orbitals as follows:

m_l	2	1	0	−1	−2
# electrons	1	1	1	0	0
m_s	1/2	1/2	1/2		

This electronic arrangement yields $L = 2 + 1 + 0 = 3$ due to the occupied orbitals, $S = 1/2 + 1/2 + 1/2 = 3/2$ and $J = |L - S| = 3/2$. The ground state is therefore designated as $^4F_{3/2}$. The ground states of relevant transition series and rare earth ions are listed in Tab. 2.2.

2.7.3 Spin–orbit coupling in the Russell–Saunders scheme

In this scheme, the energy splittings due to electronic interactions are larger than energy shifts caused by spin–orbit coupling. Spin–orbit coupling is treated as a perturbation on the electron–electron interactions, acting on the total angular momentum J states of a many-electron atom as illustrated in Fig. 2.12. The spin–orbit term $\vec{L} \cdot \vec{S}$ from eq. (2.112) can be expressed in terms of the operator

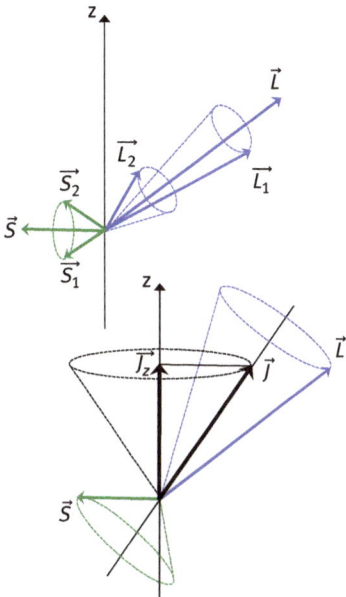

Fig. 2.11: Vector addition of angular momenta for two electrons in the Russell–Saunders scheme. Because of strong coupling, the individual orbital and spin angular momenta precess rapidly around their sums, $\vec{L}$ and $\vec{S}$, respectively, as shown on the top diagram. $\vec{L}$ and $\vec{S}$ precess slowly about the total angular momentum vector $\vec{J}$, since the coupling is weaker, as shown on the lower diagram. The projection of $\vec{J}$ onto the axis of quantization is given by $\vec{J_z}$ [3].

$$\vec{L}\cdot\vec{S} = \frac{1}{2}\left[J^2 - L^2 - S^2\right] \qquad (2.142)$$

which operates on states described by the quantum numbers J, L and S [11]. The resulting energy of a state of a given value of J is

$$E_J = \frac{\lambda}{2}[J(J+1) - L(L+1) - S(S+1)] \qquad (2.143)$$

The energy shifts due to spin–orbit coupling in a d^2 system are shown in Fig. 2.12(c): the ^3F ground state has the J values $J = 4$, 3, 2. Since the shell is less than half-full, $\lambda > 0$ and the lowest state must have $J = 2$ which is shifted relative to the ^3F state by an energy equal to

$$E_2 = \frac{\lambda}{2}[2(2+1) - 3(3+1) - 1(1+1)] = -4\lambda \qquad (2.144)$$

In the same way, the energy shifts of the remaining states can be calculated from eq. (2.143). The $(2J + 1)$ degeneracy of each J state can be removed by an external magnetic field.

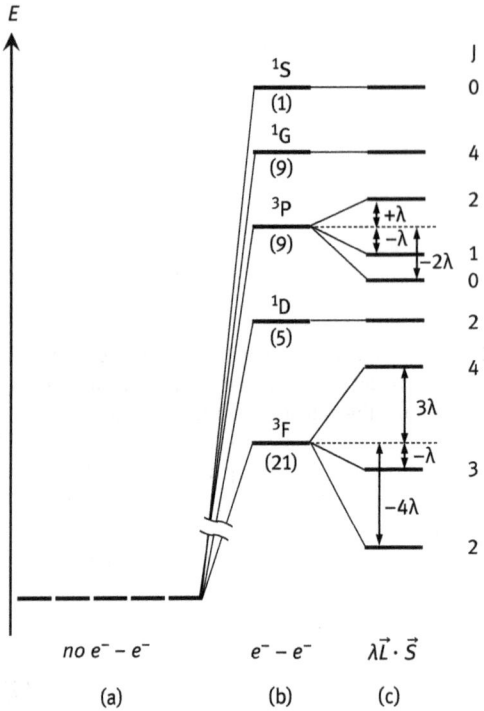

Fig. 2.12: Energy levels of a free d^2 transition series ion. (a) In the absence of electron–electron interactions, the five d-orbital states are degenerate. (b) Electronic interactions partially remove the degeneracy to give rise to five states. Numbers in parenthesis indicate the remaining degeneracy of each state. (c) Spin–orbit coupling ($\lambda > 0$) further removes the degeneracy of some of the states. Remaining degeneracy for each level is equal to $(2J+1)$. For shells that are more than half-full ($\lambda < 0$), the order of the split ^3F and ^3P levels is inverted where the state with the smaller J at a higher energy [11].

2.7.4 J–J coupling

In the case of strong spin–orbit coupling, the interaction between $\vec{L}$ and $\vec{S}$ for a particular electron is much stronger than the $\vec{L}-\vec{L}$ coupling between different electrons. We, therefore, add $\vec{L}$ and $\vec{S}$ first to give a total angular momentum

$$\vec{J}_i = \vec{L}_i + \vec{S}_i \tag{2.145}$$

for each electron and then add up the individual $\vec{J}_i$ to get the total angular momentum of the system:

$$\vec{J} = \sum_i \vec{J}_i \tag{2.146}$$

This coupling scheme is important for third-row transition metal and rare earth ions. An illustration of this can be made in Fig. 2.12, where the electron–electron repulsions are now treated as a perturbation on the spin–orbit coupling levels.

The total magnetic moment of the electron in J–J coupling is given by

$$\vec{\mu} = \vec{\mu}_l + \vec{\mu}_s = -\frac{e}{2m}\vec{L} - \frac{e}{m}\vec{S} = -\frac{e}{2m}\left(\vec{L} + 2\vec{S}\right) \tag{2.147}$$

This magnetic moment is no longer along the same direction as the total angular momentum $\vec{J} = \vec{L} + \vec{S}$, making the behavior of $\vec{\mu}$ quite complicated.

2.7.5 Quenching of the orbital angular momentum

Experiments carried out on ions of the 3d series indicate that their magnetic properties behave as though the only contribution comes from the spin magnetic moment. In other words, $\vec{J} = \vec{S}$ implying that $\vec{L} \rightarrow 0$. This *quenching* of the orbital angular momentum occurs when the ion is bound in a lattice, acted upon by a crystal field produced by the surrounding ligands. We will discuss crystal fields in a later chapter where we will see how the ligands surrounding the metal ion introduce an electrostatic potential that is dependent on the geometric arrangement of the ligands and is experienced by the electrons of the metal ion. Here it is sufficient to say that the potential experienced by an electron in such a case will no longer be a central potential, which implies that L_z will no longer be a constant of the motion. This means that L_z can be averaged out to zero:

$$\langle L_z \rangle = \psi^* L_z \psi d\tau = 0 \tag{2.148}$$

which corresponds to quenching of $\vec{L}$. When an external magnetic field is applied along the z-axis, the orbital contribution to the magnetic moment is proportional to $\langle L_z \rangle$ which means that the magnetic moment is also quenched.

The electronic magnetic moment, in this case, becomes equal to $\mu = g_e\beta\sqrt{J(J+1)} \approx g_e\beta\sqrt{S(S+1)}$ where $g_e = 2.0023$. Using the example of Cr^{3+} from above, we can calculate $g_e\beta\sqrt{J(J+1)} = 0.77\beta$ and $g_e\beta\sqrt{S(S+1)} = 3.87\beta$. For Cr^{3+} complexes and compounds, the experimentally measured value is usually $\mu = 3.8\beta$, indicating that the spin-only calculation is correct and verifying that the orbital angular momentum is quenched. A comparison of calculated and experimentally determined values of μ for transition series and rare earth ions is given in Tabs. 2.2 and 2.3. It is clear that the spin-only value is appropriate for transition series ions. Of course, deviations from the spin-only value do occur and they are due to the existence of spin–orbit coupling when there is only partial quenching of the orbital angular momentum. One such example is Co^{2+} as given in Tab. 2.2. The total angular momentum has to be considered in the case of rare earth ions where the spin–orbit coupling is strong.

The tabulated μ values become extremely relevant when analyzing EPR spectra as will be seen in later chapters.

Tab. 2.2: Tabulation of calculated and room temperature experimental magnetic moments of ions in the 3d transition metal series. The magnetic moment in the fourth column $\mu = [4S(S+1) + L(L+1)]^{1/2}\beta$ was calculated using the total angular momentum J and the moment in the last column from the spin-only expression $\mu = [4S(S+1)]^{1/2}\beta$ [12].

Ion	Orbital	State	μ/β (calc. J)	μ/β (exp.)	μ/β (calc. J = S)
Ti^{3+}, V^{4+}	d^1	$^2D_{3/2}$	3.00	1.8	1.73
V^{3+}	d^2	$^3F_2$	4.47	2.8	2.83
Cr^{3+}, V^{2+}	d^3	$^4F_{3/2}$	5.20	3.8	3.87
Mn^{3+}, Cr^{2+}	d^4	$^5D_0$	5.48	4.9	4.90
Fe^{3+}, Mn^{2+}	d^5	$^6S_{5/2}$	5.92	5.9	5.92
Fe^{2+}	d^6	$^5D_4$	5.48	5.4	4.90
Co^{2+}	d^7	$^4F_{9/2}$	5.20	4.8	3.87
Ni^{2+}	d^8	$^3F_4$	4.47	3.2	2.83
Cu^{2+}	d^9	$^2D_{5/2}$	3.00	1.9	1.73

2.8 The Zeeman effect

The Zeeman effect describes the interaction of an external magnetic field with the magnetic moments of an atom. This interaction causes a splitting of the electronic energy levels and forms the foundation of EPR. The simplest case involves the interaction of the magnetic field with the spin magnetic moment that we have briefly explored in Chapter 1 and in Section 2.4. This case applies to many transition series ions since contribution from the orbital angular momentum is often negligible. The result of the interaction is the splitting of the degenerate spin states as illustrated in Fig. 2.6. Transitions between these states give rise to EPR signals which will be discussed in detail in the next chapter.

The more general Zeeman effect arises when the orbital angular momentum is not ignored. We can distinguish three cases depending on how the external magnetic field $\vec{B}$ compares with the internal field $\vec{B}_{so}$ due to spin–orbit coupling given by eq. (2.112). The most relevant case for EPR is the weak-field Zeeman effect where $\vec{B} \ll \vec{B}_{so}$ and the fine structure must be considered first. The angular momentum that is important now is the total angular momentum vector $\vec{J}$. The interaction Hamiltonian in this case involves the total magnetic moment $\vec{\mu}$ as shown in Section 2.7.4 yielding the result

$$\mathcal{H}_{\text{Zeeman}} = -\vec{\mu} \cdot \vec{B} = -(\vec{\mu}_l + \vec{\mu}_s) \cdot \vec{B} = \frac{e}{2m}\left(\vec{L} + 2\vec{S}\right) \cdot \vec{B} \qquad (2.149)$$

Tab. 2.3: Tabulation of the magnetic moments of ions in the 4f rare earth series [13].

Ion	Orbital	State	μ/β (calc.)	μ/β (exp.)
La^{3+}	f^0	$^1S_0$	0	0
Ce^{3+}	f^1	$^2F_{5/2}$	2.54	2.4
Pr^{3+}	f^2	$^3H_4$	3.58	3.5
Nd^{3+}	f^3	$^4I_{9/2}$	3.62	3.5
Pm^{3+}	f^4	$^5I_4$	2.68	–
Sm^{3+}	f^5	$^6H_{5/2}$	0.84	1.5
Eu^{3+}	f^6	$^7F_0$	0	3.4
Gd^{3+}	f^7	$^8S_{7/2}$	7.94	8.0
Tb^{3+}	f^8	$^7F_6$	9.72	9.5
Dy^{3+}	f^9	$^6H_{15/2}$	10.63	10.6
Ho^{3+}	f^{10}	$^5I_8$	10.60	10.4
Er^{3+}	f^{11}	$^4I_{15/2}$	9.59	9.5
Tm^{3+}	f^{12}	$^3H_6$	7.57	7.3
Yb^{3+}	f^{13}	$^2F_{7/2}$	4.54	4.5

This Hamiltonian causes a splitting of the energy states relative to their zero-field value. $\mathcal{H}_{\text{Zeeman}}$ is treated as a perturbation on fine structure and using first-order perturbation theory we get the Zeeman correction to the energy to be [1]

$$E^1_{\text{Zeeman}} = \langle m_j | \mathcal{H}_{\text{Zeeman}} | m_j \rangle = \frac{e}{2m} \vec{B} \cdot \langle \vec{L} + 2\vec{S} \rangle$$

$$= \frac{e}{2m} \vec{B} \cdot \left[1 + \frac{J(J+1) + S(S+1) - L(L+1)}{2J(J+1)} \right] \langle \vec{J} \rangle \tag{2.150}$$

which for $\vec{B}$ parallel to the z-axis leads to an expression of the energy splitting of the m_j levels

$$\Delta E = g_J \beta B m_j \tag{2.151}$$

where g_J is the Landé g-factor[11]

$$g_J = 1 + \frac{J(J+1) + S(S+1) - L(L+1)}{2J(J+1)} \tag{2.152}$$

This g-factor takes values between the orbital ($g_l = 1$) and spin ($g_e = 2.0023$ for a free electron) values. We can see that inside an external field B, each degenerate energy level will be split into $2(J+1)$ levels as illustrated in Fig. 2.13 for a $3d^2$ ion.

---

11 "g-factor" is more commonly used than "g-value" here but the two terms are interchangeable.

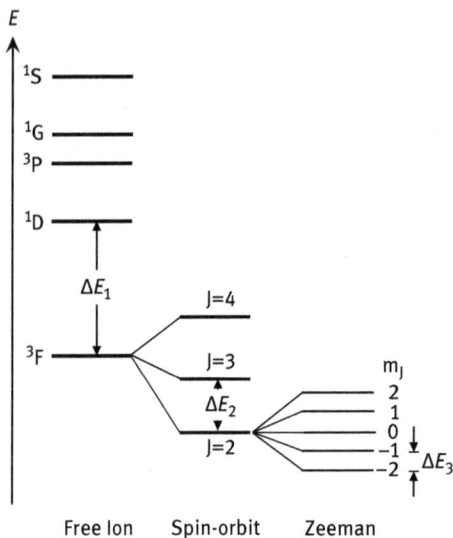

Fig. 2.13: Comparison of the energy splittings for a $3d^2$ ion in the presence of an applied field. The various splittings are not drawn to scale. Typical energies: $\Delta E_1 \sim 10^4$ cm^{-1}, $\Delta E_2 \sim 10^2$ cm^{-1}, $\Delta E_3 \sim$ 1 cm^{-1} [7].

2.9 Nuclear hyperfine interaction

Certain nuclei possess nonzero spin I which gives rise to a nuclear magnetic moment. Such nuclei interact with electrons by generating local magnetic fields at the electron that add or subtract to the external magnetic field that causes the Zeeman effect. When an electron interacts with its own nucleus, the interaction is called *hyperfine interaction* and when it interacts with nuclei from the surrounding ligands, the interaction is called *superhyperfine* interaction. The nuclear spin of some common isotopes that could contribute to an EPR signal is given in Tab. 2.4.

Tab. 2.4: Nuclear spin values of common isotopes of select elements.

Isotope	^{1}H	^{13}C	^{14}N	^{15}N	^{17}O	^{55}Mn	^{59}Co	^{63,65}Cu	^{140}Ce	^{141}Pr
I	1/2	1/2	1	1/2	5/2	5/2	7/2	3/2	0	5/2

The magnetic moment of a nucleus is given by

$$\vec{\mu}_n = \frac{g_n e}{2m_n}\vec{I} = g_n \beta_n \vec{I} \tag{2.153}$$

where g_n is the nuclear g-factor (characteristic of the particular nucleus), β_n is the nuclear magneton and $\vec{I}$ is the nuclear spin angular momentum vector. The nuclear magneton is defined as

$$\beta_n = \frac{e\hbar}{2m_p} = 5.051 \times 10^{-27} \text{J/T} \tag{2.154}$$

where m_p is the mass of the proton. Since the nucleus is much more massive than the electron, β_n is about three orders of magnitude smaller than the Bohr magneton making the nuclear magnetic moment much weaker than the electronic magnetic moment. The hyperfine interaction can be anisotropic since it originates in the interaction between the magnetic moments of the nucleus and the electron. For magnetic moments separated by a distance r, the classical dipolar interaction energy is given by [14]

$$E_{\text{dipolar}} = \frac{\mu_0}{4\pi} \frac{(1 - 3\cos^2\theta)}{r^3} \mu_{nz}\mu_{ez} \tag{2.155}$$

where μ_{nz} and μ_{ez} are the projections of the nuclear and electronic magnetic moments onto the z-axis, respectively, and θ is the angle between the line joining the two magnetic moments and the z-axis. This classical expression works well when the interacting magnetic moments are point dipoles. In the case of the electron, this approach is not adequate since the electron is described by a probability distribution extending over a volume element. For example, an s-orbital is spherical in shape and centered at the nucleus. Since the electron can exist anywhere on this sphere, we need to average the interaction over all polar and azimuthal angles θ and ϕ. This leads to an average value of $\cos^2\theta$ equal to

$$\langle \cos^2\theta \rangle = \frac{\int_0^{2\pi} \int_0^{\pi} \cos^2\theta \sin\theta d\theta d\varphi}{\int_0^{2\pi} \int_0^{\pi} \sin\theta d\theta d\varphi} = \frac{1}{3} \tag{2.156}$$

making $E_{\text{dipolar}} \to 0$. Consequently, a quantum mechanical approach is needed to fully explain hyperfine interactions.

Quantum mechanically, the external magnetic field $\vec{B}$ defines an axis of quantization relative to which we can take the projections of the electronic and nuclear magnetic moments as shown in Fig. 2.14 for the case of the hydrogen atom. The nuclear spin I follows quantization rules similar to those established for S. Relative to an axis of quantization, the nuclear spin angular momentum can adopt $(2I + 1)$ orientations determined by the value of the quantum number $m_I = -I, -I+1, \ldots, 0, \ldots, I-1, I$.

The local magnetic field at the electron due to the proton in the hydrogen atom adds or subtracts to the total magnetic field experienced by the electron. This leads to an interaction energy that causes a splitting of the Zeeman energy levels of the electron. The magnitude of this hyperfine splitting depends on the strength of the nuclear magnetic moment and the distance between the nucleus and the electron.

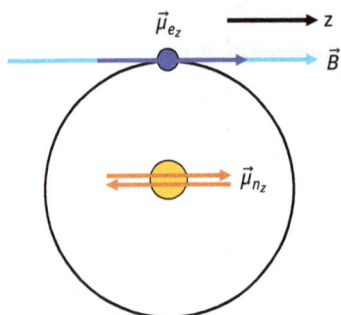

Fig. 2.14: Interaction of the magnetic moments of the electron and the proton in the hydrogen atom. The proton has a nuclear spin $I = 1/2$ with $m_I = \pm 1/2$ giving two possible orientations of the projection of $\vec{\mu}_n$ relative to the z-axis defined by the applied magnetic field $\vec{B}$.

It appears as a perturbation on the Zeeman levels. The energy separation of these new hyperfine levels is given in terms of the hyperfine constant A as shown in Fig. 2.15. This constant is an experimentally determined quantity in EPR that helps us identify the ions contributing to the EPR signals from a particular system and will be further explored in the next chapter.

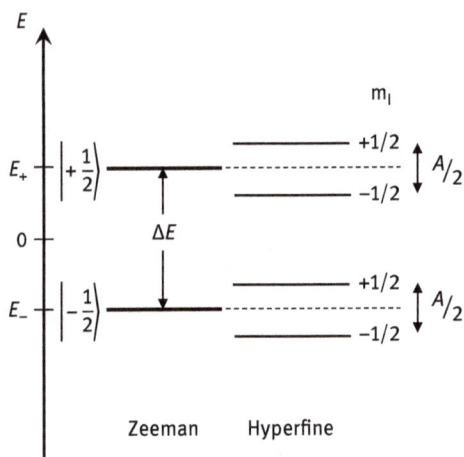

Fig. 2.15: Zeeman and hyperfine energy splittings for an $|m_s\rangle = |\pm 1/2\rangle$ state. Each m_s level is shifted by an amount $\pm A/4$, where A is the hyperfine constant, a characteristic quantity of a particular ion.

In general, for energy levels described by m_S and m_I and for the case where the hyperfine constant is small ($A \ll g\beta B$), the energy of the hyperfine levels is given by

$$E = g\beta m_S B + A m_S m_I \qquad (2.157)$$

where A is given in terms of the Planck constant h. The electronic transitions that are allowed between these energy levels will be explored in the next chapter within the context of hyperfine EPR signals.

References

[1]   D. J. Griffiths, Introduction to Quantum Mechanics (2nd ed.), Pearson Prentice Hall, 2005.
[2]   J. S. Townsend, Quantum Physics, University Science Books, 2010.
[3]   R. Eisberg and R. Resnick, Quantum Physics of Atoms, Molecules, Solids, Nuclei and Particles, John Wiley & Sons, 1974.
[4]   C. Cohen-Tannoudji, B. Diu and F. Laloe, Quantum Mechanics, vol. 1, New York: Wiley, 1977.
[5]   H. A. Bethe and R. Jackiw, Intermediate Quantum Mechanics (3rd ed.), Menlo Park, California: The Benjamin/Cummings Publishing Company, 1986.
[6]   R. Scherrer, Quantum Mechanics: An Accessible Introduction, Boston: Pearson, 2006.
[7]   J. W. Orton, Electron Paramagnetic Resonance: An Introduction to Transition Group Ions in Crystals, New York: Gordon and Breach Science Publishers Inc., 1969.
[8]   G. Palmer, "Electron Paramagnetic Resonance of Metalloproteins," In Physical Methods in Bioinorganic Chemistry – Spectroscopy and Magnetism (ed. L. Que, Jr.), Sausalito, University Science Books, 2000, pp. 121–185.
[9]   E. I. Solomon and A. B. P. Lever, Inorganic Electronic Structure and Spectroscopy (vol. I), New York: John Wiley & Sons, 1999.
[10] D. Jiles, Introduction to Magnetism and Magnetic Materials, Boca Raton, FL: CRC Press, 2016.
[11] R. S. Drago, Physical Methods for Chemists (2nd ed.), Gainesville: Surfside Scientific Publishers, 1992.
[12] F. E. Mabbs and D. J. Machin, Magnetism and Transition Metal Complexes, London: Chapman and Hall, 1973.
[13] G. Burns, Solid State Physics, Orlando, FL: Academic Press, 1985.
[14] J. A. Weil, J. R. Bolton and J. E. Wertz, Electron Paramagnetic Resonance: Elementary Theory and Practical Applications, New York: Wiley-Interscience, 1994.

3 EPR fundamentals

The quantum mechanical formalism presented in the previous chapter is applied to a detailed discussion of electron paramagnetic resonance (EPR) fundamentals in this and subsequent chapters. Our discussion in this chapter will expand on the basics of EPR presented in Chapter 1 to show how EPR signals from transition series and rare earth ions are produced and interpreted. We will also be looking at the factors that influence the shape and number of EPR lines. The discussion in this chapter covers a broad range of topics to give the reader a more complete outline of EPR theory, sufficient enough for an elementary understanding of the technique. Some of the topics presented will be further expanded in later chapters for a deeper elucidation of the material.

3.1 Magnetic resonance

The behavior of electrons and atoms inside magnetic fields is best explained using quantum mechanical arguments. The classical approach, which we used in Chapter 1 to introduce a phenomenological explanation of EPR, is limited compared to the quantum mechanical approach. It is, however, useful in allowing us to understand the behavior of single electrons and assemblies of electrons inside magnetic fields. We have seen that we can use this approach to draw conclusions about the relative orientations of the static and microwave fields that give rise to the EPR phenomenon, but we can also use it to understand relaxation processes and the lineshapes of EPR signals. The classical approach combined with quantum mechanics provides us with a more comprehensive understanding of EPR.

3.1.1 Phenomenological approach

We saw in Chapter 2 that when a magnetic moment associated with an angular momentum is placed inside an external magnetic field, the torque experienced by the magnetic moment leads to a precession of the angular momentum about the external magnetic field at the Larmor frequency. We will extend this discussion to show why an oscillating magnetic field that is orthogonal to the external magnetic field is necessary to produce the phenomenon of EPR as stipulated in Chapter 1. Let us again consider an external static magnetic field $\vec{B}_0$ directed along the z-axis interacting with a magnetic moment $\vec{\mu}$ associated with an angular momentum $\vec{L}$. This could be either the orbital angular momentum or the spin angular momentum, and the discussion works for both types. We know that

https://doi.org/10.1515/9783110417562-003

$$\vec{\mu} = -\frac{e}{2m}\vec{L} = -g\frac{\beta}{\hbar}\vec{L} = -\gamma\vec{L} \tag{3.1}$$

where $\gamma = g\beta/\hbar$ is the gyromagnetic ratio. We have seen that if the system is left alone, $\vec{\mu}$ will line up with $\vec{B}_0$. If, however, $\vec{\mu}$ is moved away from $\vec{B}_0$ so that it makes an angle θ with $\vec{B}_0$, it will experience a torque

$$\vec{\tau} = \vec{\mu} \times \vec{B}_0 \tag{3.2}$$

According to Newton's second law, the torque is also related to the rate of change of $\vec{L}$

$$\vec{\tau} = \frac{d\vec{L}}{dt} \tag{3.3}$$

Combining these three expressions leads to

$$\frac{d\vec{\mu}}{dt} = -\gamma\vec{\mu} \times \vec{B}_0 \tag{3.4}$$

We can write this in terms of the components (μ_x, μ_y, μ_z) relative to a coordinate system defined by the external magnetic field $\vec{B}_0 \parallel z$ to get the equations of motion

$$\frac{d\mu_x}{dt} = \gamma\mu_y B_0$$

$$\frac{d\mu_y}{dt} = -\gamma\mu_x B_0 \tag{3.5}$$

$$\frac{d\mu_z}{dt} = 0$$

The third equation is the easiest to solve yielding the solution that the projection of μ along the z-axis μ_z is time independent

$$\mu_z = \mu \cos\theta \tag{3.6}$$

For the first two equations, we can take the time derivative of each side and solve the resulting equations to get solutions for μ_x and μ_y subject to the boundary conditions of the problem:

$$\frac{d^2\mu_x}{dt^2} = \gamma\frac{d\mu_y}{dt}B_0 = -\gamma^2\mu_x B_0^2 \Rightarrow \mu_x = \mu\sin\theta\cos\omega_0 t \tag{3.7}$$

and

$$\frac{d^2\mu_y}{dt^2} = -\gamma\frac{d\mu_x}{dt}B_0 = -\gamma^2\mu_y B_0^2 \Rightarrow \mu_y = \mu\sin\theta\sin\omega_0 t \tag{3.8}$$

where $\mu \sin \theta$ is the projection of $\vec{\mu}$ onto the xy-plane and $\omega_0 = \gamma B_0$ is the Larmor frequency. $\vec{\mu}$ can now be expressed as follows:

$$\vec{\mu} = (\mu \sin \theta \cos \omega_0 t)\hat{i} + (\mu \sin \theta \sin \omega_0 t)\hat{j} + (\mu \cos \theta)\hat{k} \qquad (3.9)$$

(a)

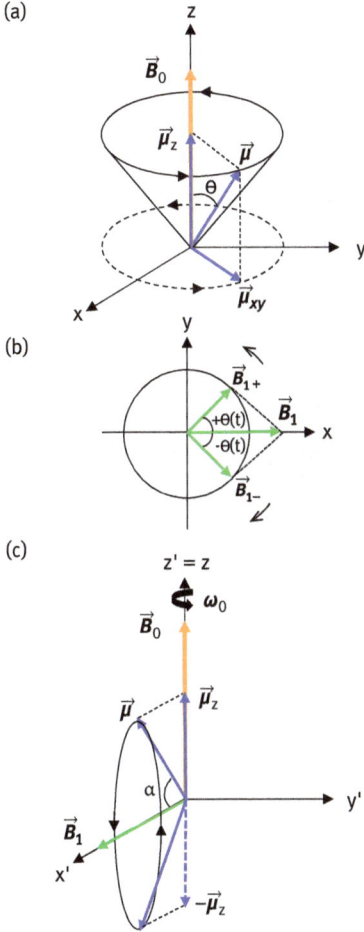

(b)

(c)

Fig. 3.1: Interaction of a magnetic moment $\vec{\mu}$ with an external static magnetic field $\vec{B}_0$ and a time-dependent magnetic field $\vec{B}_1$. (a) Precession of $\vec{\mu}$ about $\vec{B}_0$ along the z-axis at the Larmor frequency ω_0. (b) An oscillating magnetic field $\vec{B}_1$ polarized along the x-axis can be resolved into two circularly polarized components $\vec{B}_{1+}$ and $\vec{B}_{1-}$ rotating in opposite directions with an angular frequency ω: $\theta(t) = \omega t$. (c) In a coordinate frame $(x', y', z' = z)$ rotating about the z-axis with angular frequency ω_0, the magnetic moment $\vec{\mu}$ will now precess about $\vec{B}_1$ with a frequency $\omega_1 \ll \omega_0$.

This equation tells us that $\vec{\mu}$ is moving about the external field $\vec{B}_0$ with a constant angle θ, making its projection along $\vec{B}_0$ constant in time and its projection onto the plane perpendicular to $\vec{B}_0$ varying in time. This kind of motion is called precession about $\vec{B}_0$ at the Larmor frequency ω_0.

What happens to the motion of $\vec{\mu}$ when we introduce a time-dependent magnetic field in the xy-plane? Let this magnetic field be oscillating along the x-axis with a high frequency ω and have a magnitude that is much smaller than $|\vec{B}_0| = B_0$. If this field has an amplitude $2B_1$, we can express it as

$$B_x = 2B_1 \cos \omega t \tag{3.10}$$

This linearly oscillating field can be resolved into two circularly polarized components rotating in opposite directions with the same frequency ω as shown in Fig. 3.1(b) (left- and right-circular polarization). Therefore, we can express B_x as follows:

$$B_x = 2B_1 \cos \omega t = B_1 e^{i\omega t} + B_1 e^{-i\omega t} \tag{3.11}$$

We see that resonance is possible when one of the two rotating components of B_1 is synchronized with the rotating projection of $\vec{\mu}$ onto the xy-plane:

$$\vec{B}_1 = (B_1 \cos \omega_0 t)\hat{i} + (B_1 \sin \omega_0 t)\hat{j} \tag{3.12}$$

When this happens $\vec{\mu}$ experiences a total magnetic field

$$\vec{B} = (B_1 \cos \omega_0 t)\hat{i} + (B_1 \sin \omega_0 t)\hat{j} + B_0 \hat{k} \tag{3.13}$$

For the case when $B_1 \ll B_0$, which typically occurs in EPR where $0 < B_0 < 1T$ and $B_1 \sim 1\,\mu T$, it is easier to analyze the motion by switching to a rotating frame of reference (x', y', z') which rotates about the z-axis with frequency ω. The equation of motion relating $\vec{\mu}$ in the stationary frame to $\vec{\mu}'$ in the rotating frame is given by

$$\frac{d\vec{\mu}}{dt} = \frac{d\vec{\mu}'}{dt} + \vec{\omega} \times \vec{\mu} \tag{3.14}$$

which can be rearranged as

$$\frac{d\vec{\mu}'}{dt} = \frac{d\vec{\mu}}{dt} - \vec{\omega} \times \vec{\mu} = -\gamma\vec{\mu} \times \vec{B} + \vec{\mu} \times \vec{\omega} = -\gamma\vec{\mu} \times \left(\vec{B} - \frac{\vec{\omega}}{\gamma}\right) = -\gamma\vec{\mu} \times \left(\vec{B}_0 + \vec{B}_1 - \frac{\vec{\omega}}{\gamma}\right) \tag{3.15}$$

or

$$\frac{d\vec{\mu}'}{dt} = -\gamma\vec{\mu} \times \vec{B}_{\text{eff}} \tag{3.16}$$

This is telling us that in the rotating frame, the magnetic moment $\vec{\mu}$ is rotating about an effective field given by

$$\vec{B}_{\text{eff}} = \left(\vec{B}_0 + \vec{B}_1 - \frac{\vec{\omega}}{\gamma}\right) \tag{3.17}$$

with an angular frequency $\omega' = \gamma B_{\text{eff}}$. At resonance, $\omega = \omega_0$ and $\vec{B}_0$ is canceled out making $\vec{B}_{\text{eff}} = \vec{B}_1$. The equation of motion is now

$$\frac{d\vec{\mu}'}{dt} = -\gamma\vec{\mu} \times \vec{B}_1 \tag{3.18}$$

This means that in the rotating frame of reference at resonance $\vec{\mu}$ only sees $\vec{B}_1$ as a constant magnetic field (since $\vec{B}_0$ is effectively canceled) resulting in precession of $\vec{\mu}$ about $\vec{B}_1$ with angular frequency

$$\omega_1 = \gamma B_1 = \frac{g\beta}{\hbar} B_1 \qquad (3.19)$$

This is illustrated in Fig. 3.1(c), where $z' \parallel z$ and $\vec{B}_1 \parallel x'$, and $\vec{\mu}$ is at a constant angle α relative to $\vec{B}_1$. Since $\vec{B}_1 \ll \vec{B}_0$, the precession frequency ω_1 is much smaller than the Larmor frequency ω_0. We see now that this precession will cause the projection $\vec{\mu}_z$ of $\vec{\mu}$ along $\vec{B}_0$ to decrease to zero, reverse direction to $-\vec{\mu}_z$ and back again. This is equivalent to $\vec{\mu}$ flipping over and changing its energy of interaction with $\vec{B}_0$ (as $\vec{\mu}$ moves closer to $\vec{B}_0$ energy will be emitted, and as it moves away from $\vec{B}_0$ energy will be absorbed). This can be connected to what happens in a quantum mechanical two-level system when an electron changes its energy between the two spin states under the action of microwave radiation, as we will discuss in the next section.

3.1.2 Quantum mechanical approach

We saw in Chapter 2 that the Zeeman effect causes a splitting of the energy levels of the electron. In the simplest case of an isolated electron that is not bound to a particular ion, such as the ones in free radicals that exist on highly delocalized orbits, in the absence of an applied magnetic field, there are only two degenerate spin levels as discussed in Section 1.2.2. When the external magnetic field **B** is switched on, the energies of the spin levels split according to

$$E = \pm \frac{1}{2} g\beta B \qquad (3.20)$$

where $g = 2.0023$ is the electronic g-value. This relationship between E and B indicates that each energy level shifts from its zero field value in a linear fashion as the external magnetic field increases, as shown in Fig. 3.2. This leads to a splitting of the two spin levels with a separation ΔE that increases with increasing B. In a typical X-band EPR experiment, the sample is placed within the gap of an electromagnet and the magnetic field is swept from zero to about 1 T (which is equal to 10^4 Gauss, a magnetic field unit preferred in EPR). The energy separation ΔE leads to a difference in the populations of the two spin states according to the Boltzmann distribution, as shown in Section 1.2.1, that allows spins from the lowest state to be excited to the higher state. For the typical magnetic fields used in EPR, the splitting ΔE is of the order of wave numbers, which corresponds to wavelengths in the microwave region of the electromagnetic spectrum. Consequently, spins on the lowest energy state can absorb microwave photons of energy $E = h\nu$ and make transitions to the excited state. In a typical EPR experiment, the frequency of the microwave radiation incident on the sample is kept constant while the magnetic field is swept through the value B_0 that makes $\Delta E = h\nu$. This gives rise to the EPR resonance condition

$$h\nu = g\beta B_0 \qquad (3.21)$$

The frequency and the magnetic field are experimentally determined variables allowing us to determine the value of g which is one of the aims of EPR. Incorporating the numerical values of the constants h and β allows us to write the resonance condition in a more useful form as

$$g = \frac{0.71449 \cdot \nu(\text{GHz})}{B_0(\text{kG})} \qquad (3.22)$$

During the sweep of the magnetic field, the microwave power emitted by the microwave source is monitored. The EPR signal appears as an absorption peak in the

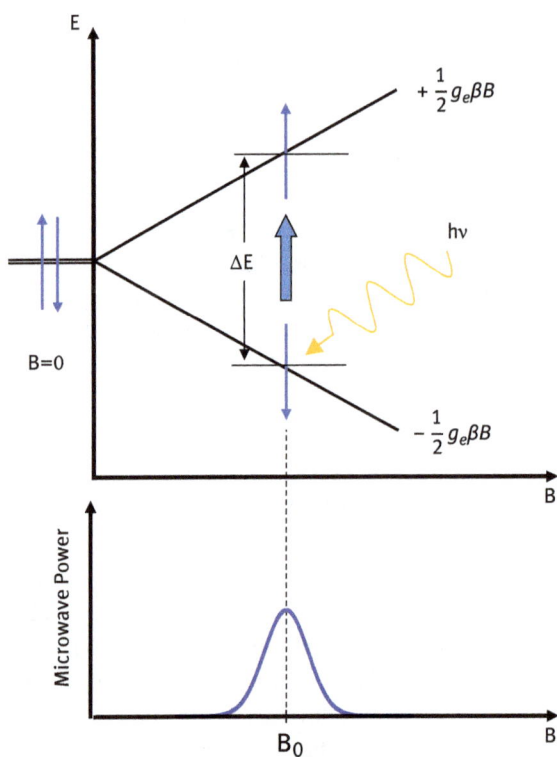

Fig. 3.2: An illustration of the Zeeman effect and how it gives rise to EPR for the spin-only case. In the absence of an external magnetic field B, the spin states are degenerate. When the magnetic field is turned on, the spin energy states split linearly with respect to the magnetic field. Microwave photons of energy $E = h\nu$ can be absorbed by electrons on the lower state and jump up to the excited state when the resonance condition $h\nu = g\beta B_0$ is satisfied. The result is the absorption of microwave energy by the sample.

source power spectrum as shown in Fig. 3.2. For sharply defined energy levels, the absorption peak would be expected to have a linewidth of zero. However, due to the Heisenberg uncertainty principle it has a finite linewidth, which can be either a Gaussian or a Lorentzian lineshape depending on characteristics of the system that will be discussed in a later section in this chapter. When the field is scanned slowly, at low microwave powers to avoid saturation and low modulation to avoid line distortion, then the integrated intensity of the absorption peak is proportional to the number of spins participating in the resonance effect. The EPR signal can be consequently used for quantitation purposes to determine the concentration of paramagnetic species by comparing the signal to that of a sample of known concentration. CuEDTA is one of the standard samples used for quantitation of frozen solution samples.

The absorption of energy by the spin system is a resonance effect and can only occur at a particular magnetic field B_0 and frequency v. Energy will not be absorbed at any other values of B as demonstrated in Fig. 3.3.

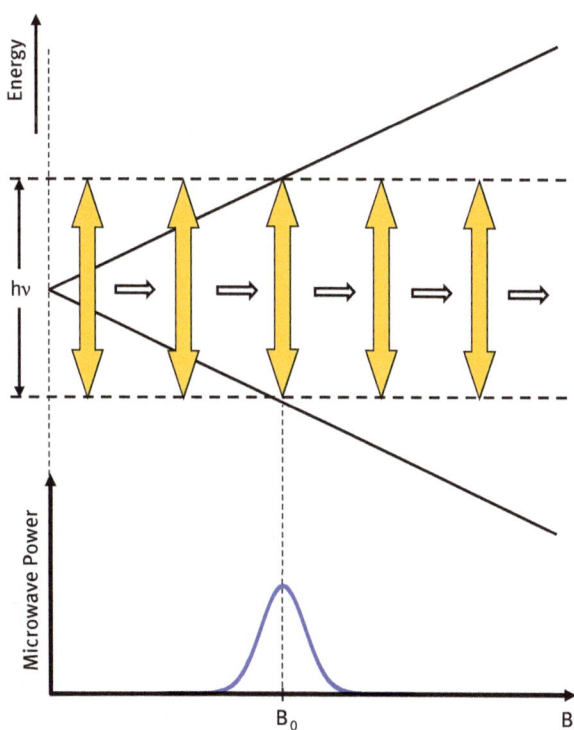

Fig. 3.3: Typical constant frequency swept magnetic field EPR experiment. The yellow double arrows represent the constant microwave energy continuously irradiating the sample. Resonance absorption of radiation only occurs at a specific value of the magnetic field B_0 when the energy of the microwave radiation matches the energy splitting of the spin levels.

Using a typical X-band frequency of ~9.5 GHz, the resonant absorption of micro-wave energy for a free radical sample ($g \approx 2.0$) occurs at a field position of ~3.390 kG. The standard free radical sample used for calibrating EPR instrumentation is DPPH (2,2-diphenyl-1-picryl hydrazyl) with one highly delocalized electron per molecule and an isotropic g-value very close to that of a free electron ($g \approx 2.00036 \pm 0.0003$). DPPH comes in a black powder form and produces a very strong EPR signal – a couple of grains is sufficient to produce a huge signal!

3.2 Relaxation and lineshapes

In order to understand relaxation mechanisms and their effect on EPR lineshapes, we need to use both of the classical and quantum resonance approaches discussed in the previous section. Instead of a single electron, we now need to consider a collection of N unpaired electrons. Typical samples studied with EPR contain very large numbers of unpaired electrons where $N > 10^{10}$. Each electron possesses a spin $\vec{S}$ with its associated magnetic moment $\vec{\mu}$. Typically, these electrons are within samples where they are separated by diamagnetic species that reduce the interactions between electrons (e.g., dipolar and exchange). We can assume that they are noninteracting electrons. In the absence of an external magnetic field, these electrons will all have the same thermal energy, which will cause their magnetic moments to point in random orientations. The magnetic properties of this spin system are described by the magnetization $\vec{M}$, which is defined as

$$\vec{M} = \sum_{i=1}^{N} \vec{\mu}_i \tag{3.23}$$

where $\vec{\mu}_i$ is the magnetic moment vector of the ith electron which has the same magnitude μ but different direction in space for each electron. In zero field, the net magnetization of the system will be equal to zero. In the presence of an external magnetic field, however, we have seen that there are two possible energy states for each electron

$$E_{\pm} = \pm \frac{1}{2} g_e \beta B_0 \tag{3.24}$$

3.2.1 Spin–lattice relaxation

The spin system is in thermal equilibrium with its environment. This environment is referred to as the *lattice*, which acts as a thermal reservoir that transfers energy in and out of the spin system. If the thermal energy $E_T = k_B T$ is of the same order of magnitude as the energy separation $\Delta E = E_+ - E_- = g_e \beta B$ of the two levels, then thermal excitation will promote spins from the ground level to the excited level giving it a finite spin population. The interaction with the lattice creates a conduit for the transfer of energy away from the spin system. We call the interaction between the spin system and the lattice the *spin–lattice* interaction that allows the spins on the excited level to transfer their energy to the lattice and drop, or relax, down to the ground level. This process is called *spin–lattice relaxation* and causes a decrease in the population of the excited level. The rate at which the energy is transferred to the lattice is given by $1/T_1$ where T_1 is the *spin–lattice relaxation time*. If the transfer of energy is very fast, T_1 is very short and the spins spend a very brief time on the excited level before dropping down to the ground level. In other words, the lifetime of the excited level is very short. We will see in a later section in this chapter that this must be avoided since it leads to a broadening of the EPR signal (lifetime broadening). Lowering the temperature of the spin system is the most direct way to slow down spin–lattice relaxation, increase T_1 and sharpen the EPR signal. Of course, if T_1 becomes too large, this can lead to saturation and disappearance of the EPR signal. In this case, the microwave power must decrease to avoid saturation as discussed below.

For such a system, the population of the excited state will depend on the temperature of the system and the energy separation of the two states according to the Boltzmann distribution:

$$N_+ = N_- e^{-\frac{g_e \beta B_0}{k_B T}} \tag{3.25}$$

where N_+ and N_- are the populations of the excited and ground states, respectively ($N_+ + N_- = N$). This expression indicates that at the limits of very high fields or very low temperatures, the population of the excited state $N_+ \to 0$. At room temperature and fields of less than 1 T, which is typical of EPR experiments carried out up to Q-band (~34 GHz), the magnetic energy is much smaller than the thermal energy: $g_e \beta B_0 \ll k_B T$. (For example, $g_e \beta B_0 = 6.121 \times 10^{-24}$ J when $B_0 = 3.3$ kG, the field where a $g_e = 2.0$ signal typically occurs at X-band frequencies and $k_B T = 4.1 \times 10^{-21}$ J at 295 K.) This inequality allows us to do a binomial expansion of the exponential to get the approximation

$$N_+ \approx N_- \left(1 - \frac{g_e \beta B_0}{k_B T} \right) = N_- - N_- \frac{g_e \beta B_0}{k_B T} \tag{3.26}$$

This expression tells us that at thermal equilibrium at room temperature and small fields, there will be a small difference between the spin populations of the two states. This difference will become even larger as we lower the temperature of the system with the majority of the spins in the ground state and fewer spins on the excited state. This situation is illustrated in Fig. 3.4(a), where we can see that since there are more spins on the ground state, there is a net spin pointing down (along the z-axis) giving rise to a net magnetization $\vec{M} = \vec{M}_0$ for the system pointing in the same direction as the magnetic field $\vec{B}_0$.

Preserving thermal equilibrium with an excess number of spins on the ground level is essential in maintaining resonance absorption that gives rise to EPR. Low microwave powers avoid quick saturation of the two levels but saturation can occur even at low microwave powers if spin–lattice relaxation is too slow. When continuous microwave radiation with energy $h\nu = \Delta E = E_+ - E_- = g\beta B$ interacts with the spins on both levels, it will cause stimulated absorption (α) and emission (ε) of radiation. At relatively low microwave powers and efficient spin–lattice relaxation, thermal equilibrium is maintained making the α process more effective than the ε process, allowing the observation of an EPR signal (Fig. 3.4(b)). The net magnetization of the system is still along the z-axis ($\vec{M}_z$) but it is now smaller than $\vec{M}_0$ since there are fewer spins on the ground level. In this case, $\vec{M}$ is tilted away from the z-axis and transverse components are also present. In the case of high microwave powers and/or inefficient spin–lattice relaxation, the populations of the two levels will be equalized leading to saturation and the disappearance of the EPR signal. Since the spin populations of the two levels are now equalized, $\vec{M}_z$ goes to zero indicating that $\vec{M}$ is tilted even further away from the z-axis and now lies entirely in the xy-plane (Fig. 3.4(c)) [1]. Return to equilibrium can be accomplished through relaxation when the microwave power is reduced. This is an exponential process controlled by T_1 where $\vec{M}_z$ increases from zero back to the value $\vec{M}_0$. Because of its effect on $\vec{M}_z$, the spin–lattice relaxation time T_1 is also called the *longitudinal relaxation time*. The time dependence of $\vec{M}_z$ as it approaches the value $\vec{M}_0$ under the influence of T_1 is given by

$$\frac{dM_z}{dt} = -\frac{M_z - M_0}{T_1} \tag{3.27}$$

3.2.2 Spin–spin relaxation

In addition to spin–lattice relaxation that involves relaxation along the z-axis, another relaxation process is present when we allow the spins to interact with one another, that is, allow energy exchanges between the spins. This interaction is called *spin–spin interaction* and is present when $\vec{M}$ is tilted away from the z-axis. It causes the $\vec{M}_x$ and $\vec{M}_y$ components of $\vec{M}$ to go to zero and is consequently referred to as *transverse relaxation*. The two components return to their equilibrium value of zero within a characteristic time called the *transverse relaxation time* T_2. This happens when the individual moments that make up $\vec{M}$ lose phase coherence during the precession of $\vec{M}$ about $\vec{B}_0$ (see next Section 3.2.3). All the individual magnetic moments are initially precessing in phase but local variations of the magnetic field and shifts in the effective g-value throughout the sample can cause different spins to precess with slightly different Larmor frequencies since $\omega_0 = g\beta B_0/\hbar$. These local variations of the magnetic field at a particular spin site are due to the magnetic moments of

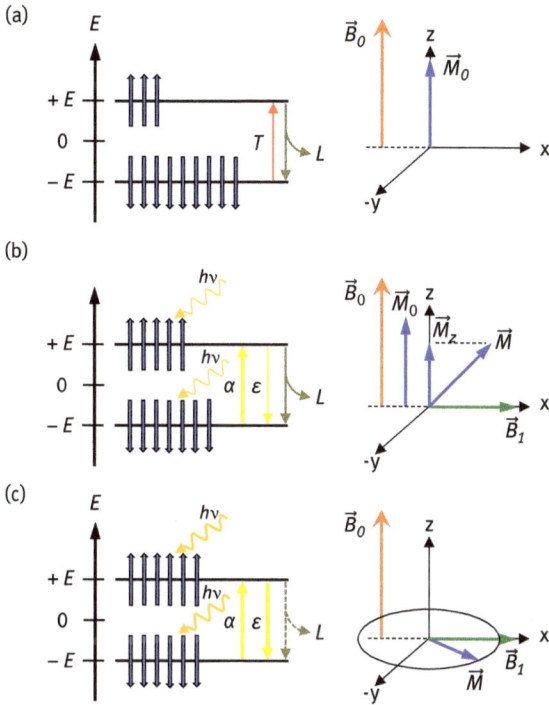

Fig. 3.4: A system of N spin magnetic moments inside an external magnetic field $\vec{B}_0$. (a) When the energy separation $\Delta E = E_+ - E_- = g\beta B$ of the two levels is of the order of the thermal energy $E_T = k_B T$, thermal excitation (T) will populate the excited level by promoting spins up from the ground level. Spin–lattice relaxation (L) transfers energy from the excited spins to the lattice maintaining a dynamic equilibrium where the number N_- on the ground level is larger than the number on the excited level, yielding an equilibrium magnetization $\vec{M}_0$ in the direction of $\vec{B}_0$. (b) During resonance, microwave radiation interacts with both levels of the spin system causing stimulated absorption (α) and emission (ε) of the radiation by the spins. Spin–lattice relaxation maintains thermal equilibrium between the two levels, allowing the absorption process to continue. $\vec{M}$ is no longer parallel to the z-axis: $\vec{M}_z < \vec{M}_0$. (c) Saturation occurs when the populations of the two levels are equalized. The rates of stimulated absorption and emission of radiation are equal. This occurs at either large microwave powers or when the spin–lattice relaxation time is long, leading to the disappearance of the EPR signal. $\vec{M}$ is now in the xy-plane and $\vec{M}_z \rightarrow 0$.

the other spins that add or subtract to $\vec{B}_0$ to give rise to an effective field experienced by the spin. Since the Larmor frequency is proportional to both the effective magnetic field and g, the spins at different sites will precess at slightly different frequencies, leading to loss of phase coherence between the spins. This process, unlike spin–lattice relaxation, does not involve any transfer of energy between the spin system and the lattice. The time dependence of $\vec{M}_x$ and $\vec{M}_y$ as they decay to their equilibrium value of zero is given by

$$\frac{dM_x}{dt} = -\frac{M_x}{T_2} \tag{3.28}$$

and

$$\frac{dM_y}{dt} = -\frac{M_y}{T_2} \tag{3.29}$$

3.2.3 The Bloch equations

The effects of relaxation on a two-level system can be explained by going back to the precession of a magnetic moment about an external magnetic field. The same discussion that applies to $\vec{\mu}$ can be extended to the magnetization $\vec{M}$: instead of a single magnetic moment precessing about $\vec{B}_0$ we now have the magnetization of the spin ensemble precessing about $\vec{B}_0$. We modify the equation of motion for the precession of $\vec{\mu}$ about $\vec{B}_0$ (eq. (3.4)) to now become:

$$\frac{d\vec{M}}{dt} = -\gamma \vec{M} \times \vec{B}_0 \tag{3.30}$$

leading to precession of $\vec{M}$ about $\vec{B}_0$ at the Larmor frequency ω_0. When the orthogonal oscillating field

$$\vec{B}_1 = (B_1 \cos \omega t)\hat{i} + (B_1 \sin \omega t)\hat{j} \tag{3.31}$$

with frequency $\omega = \omega_0$ is applied, we can affect changes in $\vec{M}_z$ as described earlier. ω_0 then coincides with the angular frequency associated with transitions between the states of the two-level system. The equation of motion of $\vec{M}$ can be transformed in a frame of reference that is rotating about the z-axis at the Larmor frequency. The equation of motion is now written as (from p. 63)

$$\frac{d\vec{M}}{dt} = -\gamma \vec{M} \times \vec{B}_1 \tag{3.32}$$

indicating that $\vec{M}$ is now precessing about $\vec{B}_1$ with frequency $\omega_1 = \gamma B_1$ just as $\vec{\mu}$ did in our earlier discussion. The resulting equations of motion for the components of $\vec{M}$ in the presence of both $\vec{B}_0$ and $\vec{B}_1$ were derived by Bloch (the Bloch equations) with the incorporation of relaxation processes to yield [2]

$$\frac{dM_x}{dt} = -\gamma (M_y B_0 - M_z B_1 \sin \omega t) - \frac{M_x}{T_2} \tag{3.33}$$

$$\frac{dM_y}{dt} = -\gamma (-M_x B_0 + M_z B_1 \sin \omega t) - \frac{M_y}{T_2} \tag{3.34}$$

$$\frac{dM_z}{dt} = = -\gamma(M_xB_1 \sin \omega t - M_yB_1 \cos \omega t) - \frac{M_z - M_0}{T_1} \tag{3.35}$$

Transforming to a rotating coordinate system (u, v, z) with u parallel to $\vec{B}_1$, v at $90°$ to $\vec{B}_1$ in the same plane as u, and z parallel to $\vec{B}_0$ yields [3]

$$\frac{du}{dt} = -(\omega - \omega_0)v - \frac{u}{T_2} \tag{3.36}$$

$$\frac{dv}{dt} = (\omega - \omega_0)v - \frac{v}{T_2} + \gamma B_1 M_z \tag{3.37}$$

$$\frac{dM_z}{dt} = -\gamma B_1 v - - \frac{M_z - M_0}{T_1} \tag{3.38}$$

where

$$u = M_x \cos \omega t + M_y \sin \omega t \tag{3.39}$$

$$v = M_x \sin \omega t - M_y \cos \omega t \tag{3.40}$$

The Bloch equations in the rotating frame can be simplified and solved in special cases when certain terms become negligible. One of these cases is *slow passage* through resonance where $\vec{B}_0$ changes slowly compared to relaxation times. In such a case, u, v and M_z change very slowly making the derivatives of these variables negligible compared to the actual values: $du/dt \ll u/T_2$, $dv/dt \ll v/T_2$, $dM_z/dt \ll M_z/T_1$. We can, therefore, ignore them in the Bloch equations above and solve to obtain

$$u = \frac{\gamma B_1 M_0 (\omega_0 - \omega) T_2^2}{1 + (\omega_0 - \omega)^2 T_2^2 + \gamma^2 B_1^2 T_1 T_2} \tag{3.41}$$

$$v = \frac{\gamma B_1 M_0 T_2}{1 + (\omega_0 - \omega)^2 T_2^2 + \gamma^2 B_1^2 T_1 T_2} \tag{3.42}$$

$$M_z = \frac{M_0 \left[1 + (\omega_0 - \omega)^2 T_2^2\right]}{1 + (\omega_0 - \omega)^2 T_2^2 + \gamma^2 B_1^2 T_1 T_2} \tag{3.43}$$

These equations allow us to verify arguments we made about the behavior of the two-level system in the presence of $\vec{B}_0$ and $\vec{B}_1$. The variables u and v which are due to $\vec{M}$ deviating from equilibrium have finite values only when $\vec{B}_1$ is present, indicating that $\vec{B}_1$ is responsible for driving the system away from equilibrium. They go to zero when $\vec{B}_1 \to 0$ and at the same time $M_z \to M_0$, indicating a return to equilibrium.

The term B_1^2 is proportional to the microwave power. At low microwave powers, B_1^2 is small making the term $1 + \gamma^2 B_1^2 T_1 T_2 \approx 1$ in the denominator of the expressions for u and v. These expressions now become

$$u \approx \frac{\gamma B_1 M_0 \Delta\omega T_2^2}{1 + (\Delta\omega)^2 T_2^2} \tag{3.44}$$

$$v \approx \frac{\gamma B_1 M_0 T_2}{1 + (\Delta\omega)^2 T_2^2} \tag{3.45}$$

where $\Delta\omega = (\omega_0 - \omega)$. Plots of u and v versus $T_2\Delta\omega$ are shown in Fig. 3.5. At resonance, when the frequency ω of the microwave field becomes equal to the Larmor frequency ($\omega = \omega_0 = \gamma B_0 \Rightarrow \Delta\omega = 0$), the component u goes to zero. This is the component which is rotating in phase with $\vec{B}_1$. Since it goes to zero at resonance, it does not represent absorption of microwave power but rather it is associated with dispersion. The v component is the 90° out-of-phase component relative to the $\vec{B}_1$ rotation, and the plot of v versus $T_2\Delta\omega$ indicates absorption of microwave power at resonance. We can distinguish the two components as the absorptive (u) and the dispersive (v) components.

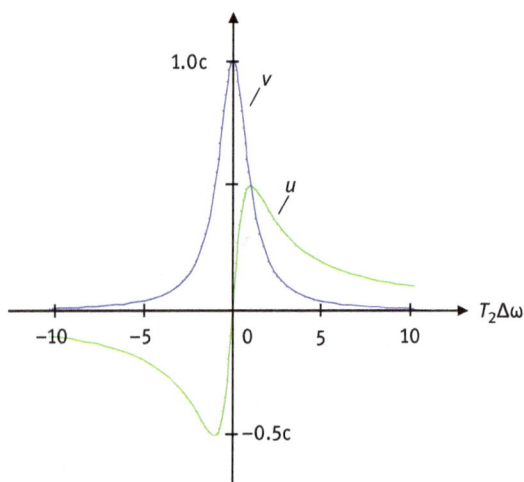

Fig. 3.5: Behavior of the rotating components u and v as a function of angular frequency. Resonance occurs when $\Delta\omega = (\omega_0 - \omega) = 0$ with the simultaneous absorption of microwave energy as reflected in the maximum displayed in the curve for the v component. The units on the vertical axis are in terms of $c = \gamma B_1 M_0 T_2$. This lineshape is a Lorentzian line.

The shape of the v curve is mathematically described by a Lorentzian line. The plot of v versus $T_2\Delta\omega$ is reproduced again in Fig. 3.6 for the limiting cases of low ($\gamma^2 B_1^2 T_1 T_2 \ll 1$) and large ($\gamma^2 B_1^2 T_1 T_2 \sim 1$) microwave powers. Higher microwave powers cause a broadening of the absorption line accompanied by a decrease in the amplitude of v and a consequent reduction of the EPR signal intensity. This corresponds to the populations of the states of the two-level system becoming equal when spin–lattice relaxation cannot maintain equilibrium. The broadening of the line is measured in terms of the full-width at half-maximum (FWHM) which, for the low microwave power case, occurs at $T_2\Delta\omega = 2$. We can consequently express the linewidth of the Lorentzian as

$$\Delta\omega_{1/2} = \frac{2}{T_2} \tag{3.46}$$

Since the EPR signal is typically recorded when the frequency of the microwave radiation is kept constant and the external field $\vec{B}$ is varied, we can express the linewidth in terms of the magnitude of $\vec{B}$ by using the resonance condition $h\nu = g\beta B$ (or $\hbar\omega = g\beta B$) and $\Delta\omega = 2\pi\Delta\nu$ to write it as

$$\Delta B_{1/2} = \frac{2\hbar}{g\beta T_2} \tag{3.47}$$

We see that the width of a Lorentzian line at low microwave powers is influenced by T_2 and not T_1: the shorter T_2 is, the broader the linewidth. Even though we do want T_2 to be short to keep the term $\gamma^2 B_1^2 T_1 T_2 \ll 1$, a too short T_2 can lead to significant line broadening and loss of signal. Consequently, T_2 must be kept within a certain range to ensure that the EPR signal is sharp enough. As an example, we can consider concentrated samples of many paramagnetic materials where T_2 can be quite fast since the paramagnetic species are strongly interacting. Diluting the sample by incorporating the paramagnetic species within a diamagnetic host weakens the interactions and lengthens T_2 to yield a reasonable ΔB.

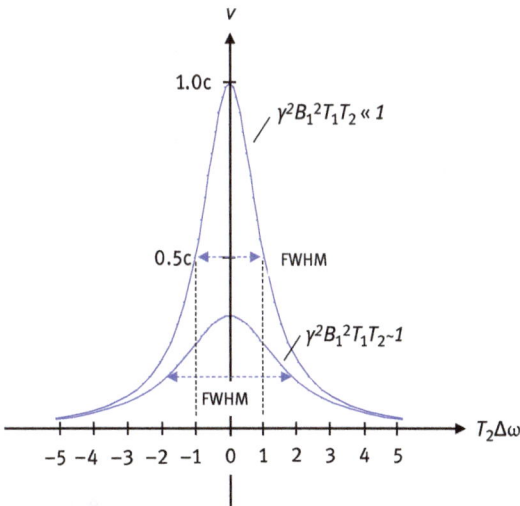

Fig. 3.6: Lorentzian absorption line plotted versus angular frequency at low and large microwave powers. The microwave power is proportional to B_1^2 and affects the $\gamma^2 B_1^2 T_1 T_2$ term in eq. (3.41). At high microwave powers saturation occurs, causing a broadening of the absorption line and a decrease of the EPR signal. The broadening is characterized by the linewidth, which is expressed in terms of the full-width at half-maximum. The units on the vertical axis are in terms of $c = \gamma B_1 M_0 T_2$.

In a typical EPR experiment, phase-sensitive detection is employed, resulting in the recording of the derivative of the EPR signal. Taking the derivative of v with respect to ω yields the expression

$$\frac{dv}{d\omega} = \frac{-2\gamma B_1 M_0 T_2^3}{\left[1 + (T_2\Delta\omega)^2\right]^2}\Delta\omega \tag{3.48}$$

which is plotted in Fig. 3.7. In this case, the linewidth is determined as the peak-to-peak separation of the turning points. From the plot, we see that the peak-to-peak separation occurs at

$$\Delta\omega_{pp} = \frac{2}{\sqrt{3}T_2} \tag{3.49}$$

or in terms of the applied magnetic field [3]

$$\Delta B_{pp} = \frac{2\hbar}{\sqrt{3}g\beta T_2} \tag{3.50}$$

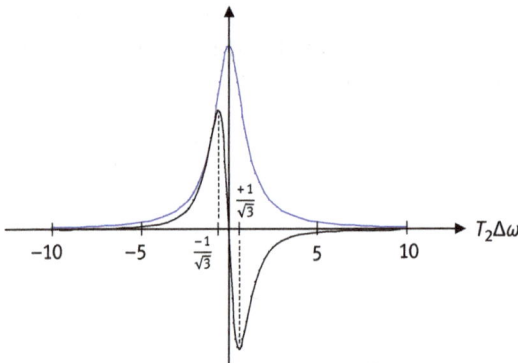

Fig. 3.7: Plot of v (blue) and $dv/d\omega$ (black) as a function of angular frequency. The separation of the two peaks of the first derivative is $T_2\Delta\omega_{pp} = 2/\sqrt{3}$ which allows the determination of the peak-to-peak linewidth.

3.3 Further discussion of relaxation mechanisms

As we have already seen, the transfer of energy away from a spin system is achieved through two types of "relaxation" mechanisms. These are *spin–lattice relaxation* when the energy is transferred to the rest of the molecule and/or crystal lattice via the emission or absorption of phonons, and *spin–spin relaxation* when the energy is transferred to other spins. Both of these relaxation mechanisms are nonradiative processes. The times associated with these processes that indicate how fast energy is transferred are the spin–lattice relaxation time T_1 and the spin–spin relaxation time T_2. We have discussed these times in the context of the Bloch equations as the longitudinal (T_1) and transverse (T_2) relaxation times. Both types of interaction affect the linewidth, which yields information on how strongly the paramagnetic ion is coupled to the lattice and, through the spin–spin interaction, bonding and structural information can be obtained. The discussion in this section will be focused on solid samples since these account for the majority of samples studied by EPR.

Spin–lattice relaxation is the mechanism that maintains the difference in populations between the two spin states that allows the observation of an EPR signal, as we have already discussed in the previous section. Without it, the incident microwave radiation would quickly equalize the populations of the two levels and no further absorption of radiation would occur, leading to saturation effects and deterioration of the signal. The origin of spin–lattice relaxation is in the thermal vibrations of the atoms or ions in the crystal lattice. These lattice vibrations cause crystal field modulations, which produce a fluctuating crystal field at the paramagnetic ion. We will see in a later section that the crystal field is an electric field, which does not directly interact with the spin magnetic moments of the electrons. The interaction with the spins is an indirect one through the spin–orbit coupling since the crystal field directly interacts with the orbital motion of the electrons. The static crystal field, therefore, indirectly influences the spin levels through spin–orbit coupling, and an oscillating crystal field can cause transitions between these levels [4]. An analogy can be drawn here with the way an oscillating microwave magnetic field causes transitions between the spin states. Since the interaction is through spin–orbit coupling, the strongest spin–lattice interaction occurs with ions that have considerable orbital contributions to their ground state wave function such as the rare earth ions. This makes the lifetime of the excited state in the rare earths and T_1 very short.

The lattice vibrations generate waves called Debye waves or phonons. The phonon is the quantum of energy of lattice vibrations in direct analogy to the photon in electromagnetic radiation. The number of phonons increases with frequency up to a maximum frequency called the Debye frequency v_D according to

$$k\theta_D = hv_D \tag{3.51}$$

where θ_D is the Debye temperature which is characteristic of the substance under study and k is the Boltzmann constant. The number of phonons of a particular frequency v is a function of the lattice temperature and increases as the temperature T increases according to the phonon occupation number N [5]:

$$N = \frac{1}{e^{hv/kT} - 1} \tag{3.52}$$

Since the spin–lattice relaxation time depends on the interaction between the spin system and these phonons, it is strongly dependent on the temperature of the system. The unpaired electrons of certain species, such as the rare earth ions that strongly couple to the lattice through spin–orbit coupling, have short relaxation times T_1 at room temperature, which leads to very broad lines. A short T_1 reduces the lifetime of the excited state leading to an increase in ΔB (see Section 3.5.1). In order to decrease the linewidth and sharpen the EPR signals, we often need to go to very low temperatures to slow down T_1. In the case of rare earth ions, we quite often need to go to temperatures below 20 K in order to observe a signal. For transition series ions where the orbital angular momentum is partially or totally quenched, T_1 is typically quite long at room temperature and consequently EPR signals can be observed at these temperatures.[1]

There are three main spin–lattice mechanisms: the direct or one-phonon process, the Raman or two-phonon process and the Orbach process. The direct process is a resonant mechanism where the spin on the excited state $|2\rangle$ with energy $E_2 = +\frac{1}{2}g\beta B$ transfers its energy to the lattice and relaxes down to the $|1\rangle$ state of energy $E_1 = -\frac{1}{2}g\beta B$ by emitting a single phonon at the resonant frequency ν_0 with energy equal to $g\beta B$. Only phonons of one particular frequency are involved in this process, which dominates at very low temperatures ($T < 4.2$ K) and is linearly dependent on T [5]. The Raman process is a two-phonon, nonresonant scattering mechanism. A phonon of frequency ν_1 is absorbed and another of frequency ν_2 is emitted with the difference $\nu_2 - \nu_1$ equal to the resonant frequency ν_0. The Raman process involves phonons from any part of the lattice vibration spectrum as long as the difference in their frequencies is equal to ν_0. It is dominant at higher temperatures ($T > 4.2$ K) where more phonons with the correct initial frequency ν_1 can be found. The Orbach process is also a two-phonon resonant process, which is actually a combination of two resonant direct process transitions. It requires an excited electronic state $|3\rangle$ above the $|2\rangle$ state. The spin, which is originally on state $|2\rangle$, absorbs a phonon of frequency ν_2 from the lattice that excites it to the $|3\rangle$ state and then returns to the ground state $|1\rangle$ by emitting a phonon of frequency ν_3. The Orbach process competes with the Raman process at higher temperatures. The three mechanisms are illustrated in Fig. 3.8.

Spin–spin relaxation transfers energy from the ion under investigation to other magnetic ions in the lattice. The primary interaction mechanism is through the rotating component of the magnetic moment of neighboring ions. We can think of the magnetic moments of the interacting ions as precessing about the external magnetic field in the manner discussed in Section 1.2.1 and illustrated in Fig. 1.7. If the rate of precession is the same for the two magnetic moments, then the perpendicular components of $\vec{\mu}$ of the neighboring electrons will produce an oscillating field at the electron and induce transitions from the excited state E_2 to the ground state E_1 and, consequently, reduce the lifetime of the excited state. This leads to an increase in the natural linewidth and broadening of the line. This type of "dipolar" broadening is independent of the temperature and the applied magnetic field. It varies as

---

1 As an example of the effect of sample temperature on T_1, consider the case of Cr(III) doped in MgO: T_1 varies from 800 ms at 1.3 K to 3.4 ms at 50 K, a change larger than two orders of magnitude [32].

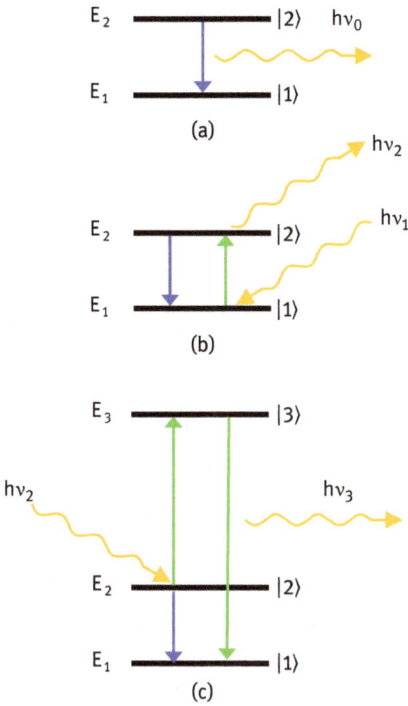

Fig. 3.8: An illustration of the three spin–lattice relaxation mechanisms. (a) Direct or one-phonon process: spin makes transition from the upper to the lower state (energies: $E_1 = -\frac{1}{2}g\beta B$ and $E_2 = +\frac{1}{2}g\beta B$) with the energy converted into a single phonon at the resonant frequency v_0. Phonon energy = $g\beta B$. (b) Raman or two-phonon process: spin absorbs phonon of frequency v_1 and gets excited to energy state E_2 before making a transition to the ground state E_1 by emitting a phonon of frequency v_2: $(v_2 - v_1) = v_0$. (c) Orbach process: two-phonon process where spin on the excited state at E_2 absorbs phonon of frequency v_2 to jump to a further excited state at E_3. Spin drops down to state at E_1 with the emission of another phonon of frequency v_3. Blue arrows indicate EPR transitions and green arrows indicate phonon transitions.

$1/r^6$, where r is the distance between the ion and its neighbors; hence, the only way to decrease this kind of broadening is to increase the distance r. This is achieved by diluting the paramagnetic species with an isomorphous diamagnetic species resulting in a magnetically dilute sample as shown in Fig. 3.9 [6]. Dilutions of a few molar percent usually produce sharp lines.

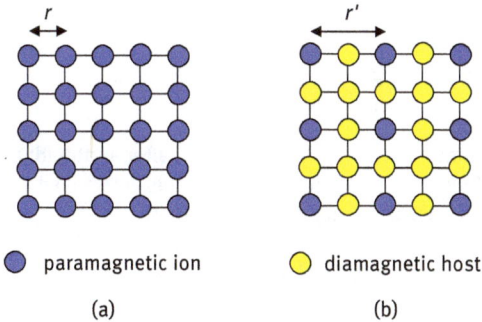

Fig. 3.9: Effect of dilution on the interionic distance r between paramagnetic species in a square lattice. (a) In a concentrated lattice, all lattice sites are occupied by paramagnetic ions. (b) Paramagnetic ions doped within a diamagnetic host. The nearest neighbor distance r' between paramagnetic species is now larger.

3.4 Experimental determination of relaxation times

Because of the different nature of the two types of relaxation processes, the experimental methods used in determining relaxation times are also different. Two examples of measurements of the relaxation times of homogeneously broadened lines using continuous wave EPR are given below:

T_1: an EPR signal is recorded under conditions of negligible saturation. With the field set at resonance, the signal is then saturated using a short microwave pulse at high power that causes the intensity to go to zero. The signal recovers at the end of the pulse following an exponential rise curve. T_1 is measured as the time it takes for the signal to reach $(1-(1/e))$ of its unsaturated intensity as shown in Fig. 3.10 [7].

T_2: directly determined by measuring the linewidth of a Lorentzian line according to

$$\frac{1}{T_2} = \frac{g\beta}{2\hbar}\Delta B_{1/2} = \frac{\sqrt{3}g\beta}{2\hbar}\Delta B_{pp} \tag{3.53}$$

The measurement must take place under nonsaturating conditions to ensure that thermal equilibrium is not disturbed. In general, it is difficult to get accurate measurement of T_2 from the lineshape since dynamic processes and inhomogeneous broadening affect the linewidth as we will see in the next section.

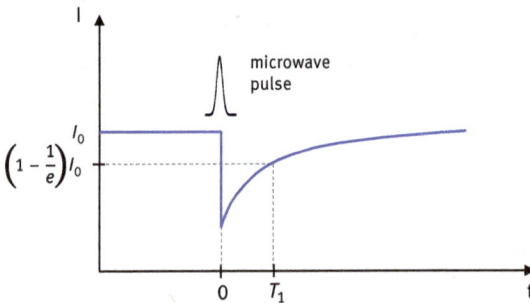

Fig. 3.10: Measurement of T_1 by monitoring the time for recovery of an EPR signal after a saturating microwave pulse is applied at $t = 0$ s. Signal intensity I versus time t is plotted. T_1 is determined at the time where the signal recovers approximately 63% of its equilibrium intensity I_0. Figure adapted from data on Ce(III) doped in $La_2Mg_3(NO_3)_{12}$ $24H_2O$ in [7]. In those experiments, T_1 was found to be equal to ~10 ms at 1.75 K and to be strongly dependent on the temperature T of the sample.

3.5 EPR lineshapes

In our discussion of the behavior of the magnetization of a spin system in the presence of $\vec{B}_0$ and $\vec{B}_1$, we saw that the magnetization equations of motion predicted that the resonance phenomenon is described by an absorption line given by a Lorentzian function. There is another common type of EPR lineshape described by a Gaussian function. Whether an EPR lineshape is a Lorentzian or Gaussian depends on the mechanism that causes the broadening: if the mechanism is due to lifetime broadening, then the lineshape is Lorentzian. The broadening mechanisms that give rise to Gaussian lineshapes are discussed below. The functions describing these lineshapes are typically plotted in terms of the magnetic field B. Following the discussion in the previous section, we can transform the Lorentzian function of the absorption versus ω given in eq. (3.45) to an expression versus B by using the resonance condition $\hbar\omega = g\beta B$. The resulting Lorentzian expression is

$$A = \frac{c}{1 + a^2(B_0 - B)^2} \tag{3.54}$$

with $c = \gamma B_1 M_0 T_2$ and $a = (\gamma T_2)^2$. This absorption function has the same maximum value c as the frequency-dependent function but is now centered at the resonant field B_0 instead of ω_0. An equivalent Gaussian function with the same resonant field can be expressed as

$$A = c' e^{-b^2(B_0 - B)^2} \tag{3.55}$$

where c' is the maximum at B_0 and b is a measure of the linewidth $\Delta B_{1/2}$: $b = 2\sqrt{\ln 2}/\Delta B_{1/2}$. The two functions are frequently written in the more convenient form [8]:

$$A = \frac{A_0}{1 + \left(\frac{B_0 - B}{\Delta B_{1/2}/2}\right)^2} \tag{3.56}$$

for the Lorentzian and

$$A = A_0 e^{-\ln 2 \left(\frac{B_0 - B}{\Delta B_{1/2}/2}\right)^2} \tag{3.57}$$

for the Gaussian. The two functions are plotted versus B in Fig. 3.11. The FWHM $\Delta B_{1/2}$ is the linewidth of the signal, which is an important experimental parameter that can provide useful information about the sample. As already mentioned, the first derivative of each function is what is typically recorded during an EPR experiment that employs phase-sensitive detection and is shown in the same figure. The experimental quantity of interest in this case is the peak-to-peak linewidth ΔB_{pp}. It can be seen from Fig. 3.11, where the two functions with the same $\Delta B_{1/2}$ are plotted, that ΔB_{pp} is slightly different for the two functions. The relationship between $\Delta B_{1/2}$ and ΔB_{pp} is given by

$$\Delta B_{1/2} = \sqrt{3}\Delta B_{pp} \quad \text{and} \quad \Delta B_{1/2} = \sqrt{2\ln 2}\Delta B_{pp} \tag{3.58}$$

for the Lorentzian and Gaussian, respectively. It should be noted that the line-shapes of real experimental signals fall somewhere in between these two cases, depending on the relative strengths of the factors that influence the line broadening. In cases where both Lorentzian and Gaussian broadening mechanisms are acting independently on the same line, the result is a convolution function of the Lorentzian and Gaussian functions [8].

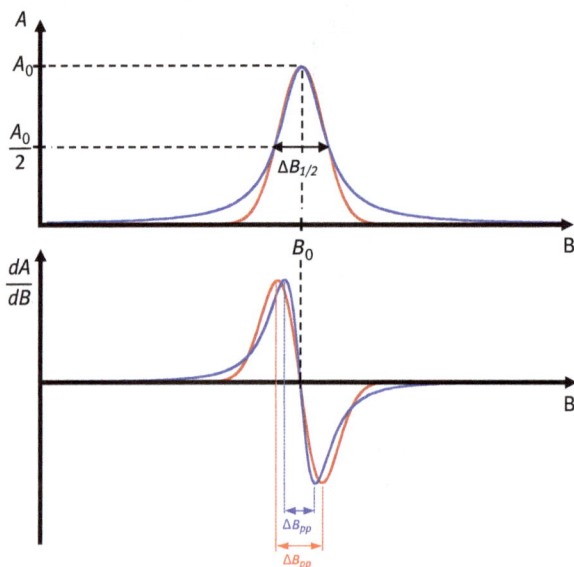

Fig. 3.11: Comparison of Lorentzian (blue) and Gaussian (red) lineshapes with the same maximum A_0 and full-width at half-maximum: the Lorentzian is sharper than the Gaussian close to B_0 but broader in the wings. The top plot is of the functions given by eqs. (3.56) and (3.57), and the bottom plot is the first derivative of the two functions. The full-width at half-maximum linewidths $\Delta B_{1/2}$ for the absorption lines and the peak-to-peak linewidths ΔB_{pp} for the derivative lines are also shown.

The EPR linewidth is an experimental quantity that yields information on the interaction between the spins of the paramagnetic species and its environment. This information is obtained via the relaxation times T_1 and T_2, which can provide information on how strongly the spin system is coupled to the lattice and other neighboring spins [9].

There are two factors that contribute to the linewidth of the EPR signal: homogeneous and inhomogeneous broadening.

3.5.1 Homogeneous broadening

This type of broadening is also referred to as "lifetime broadening" due to the finite lifetime of the excited state of the electron participating in the resonance process. When the electron absorbs microwave energy and is promoted to the excited state, it can return to the ground state (it will "relax") by transferring this energy either to the rest of the molecule/lattice or it can transfer it to other spins as we have seen in Section 3.3. Consequently, the electron spends a finite time on the excited state that leads to lifetime broadening. This is a direct consequence of the Heisenberg uncertainty principle,

$$\Delta E \cdot \Delta t \sim \hbar \tag{3.59}$$

where Δt is the mean lifetime of the excited state and ΔE is the uncertainty in the measurement of the energy of the state. Using the resonance condition $E = g\beta B$, we can rewrite the uncertainty principle in terms of the linewidth ΔB to get a relationship between ΔB and Δt:

$$g\beta\Delta B \cdot \Delta t \sim \hbar \Rightarrow \Delta B = \frac{\hbar}{g\beta\Delta t} \tag{3.60}$$

When the lifetime of the excited state is reasonably long, the spin spends more time on the excited state, making the uncertainty in both the energy of the state and the resonance field smaller. ΔB is small but finite and represents the natural width of the homogeneously broadened line. Homogeneously broadened lines have a Lorentzian lineshape. On the other hand, when the spin system is strongly coupled to the lattice/molecule, energy is transferred away fast and Δt becomes small. For such short relaxation times, the uncertainty in energy ΔE becomes large creating a spread of the magnetic field values over which resonance can occur, leading to overly broad EPR lines that might be difficult to detect. As we have already discussed in Section 3.3, this process can be controlled by going to low temperatures to slow down the spin–lattice relaxation time to make Δt longer and sharpen the EPR line by decreasing ΔB.

3.5.2 Inhomogeneous broadening

An inhomogeneously broadened EPR line is an envelope that contains a number of narrower, overlapping, homogeneously broadened lines called "spin packets" as illustrated in Fig. 3.12. Each spin packet is produced by spins in the sample that experience the same magnetic field. Most inhomogeneously broadened lines have an approximately Gaussian lineshape [10]. Some of the major causes of the displacement of the spin packets from the resonant frequency that lead to line broadening are:

Dipolar interactions: these are interactions between a particular central spin and the magnetic dipole moments of other paramagnetic ions in the sample. This time we look at the projection $\vec{\mu}_{\parallel}$ of these magnetic dipoles along the direction of the magnetic field ($\vec{\mu}_{\parallel} = \vec{\mu}_z$ when $\vec{B}$ is parallel to the z-axis as shown in Fig. 3.1) which is produced when the magnetic moment of one of these paramagnetic ions precesses about the external field. $\vec{\mu}_{\parallel}$ has a constant magnitude that interacts with the central spin, which now experiences a local field due to the $\vec{\mu}_{\parallel}$ of the magnetic moments of all of the other paramagnetic ions in the lattice. The resulting field then experienced by the central spin is equal to the external field plus the dipolar field. Since the dipolar interaction varies as $(1 - 3\cos^2\theta)$ with θ the angle between the external magnetic field and the line joining the central spin to another ion,[2] the interaction with different neighboring ions at different values of θ will shift the local fields experienced by the central spin producing small shifts in the energy levels and broadening the line [6]. If we assume a random distribution of ions in the crystal, the line will have a Gaussian shape.

Unresolved hyperfine structure: this is due to nonmagnetic ions in the lattice with large nuclear magnetic moments. Their interaction with the central ion produces small overlapping hyperfine splittings that broaden the EPR line.

Inhomogeneity of the crystal lattice: magnetic ions at symmetrically inequivalent sites will experience slightly different crystal fields. This causes a spread of the spin energy levels that also broaden the EPR line.

Inhomogeneity of the applied magnetic field: the electromagnets used in EPR experiments are designed to produce highly homogeneous magnetic fields between the pole caps. It is possible, however, to have small inhomogeneities that extend over the volume of the resonant cavity exposing parts of the sample to slightly different fields leading to line broadening.

3.5.3 Line narrowing

There are certain effects that instead of broadening an EPR line they can produce narrowing effects. One such effect is *motional narrowing* of the line which can reduce the effects of magnetic dipole broadening. This occurs in liquid samples where there is rapid motion of the molecules that averages the term $(1 - 3\cos^2\theta)$ to zero resulting in no dipolar broadening. Another effect is that of *exchange narrowing*

---

2 The potential energy of interaction between two randomly oriented magnetic dipoles $\vec{\mu}_i$ and $\vec{\mu}_j$, separated by a vector $\vec{r}$ that makes an angle θ with the z-axis, is given by $U = \left((\vec{\mu}_i \cdot \vec{\mu}_j / r^3) - (3(\vec{\mu}_i \cdot \vec{r})(\vec{\mu}_j \cdot \vec{r})/r^5)\right)$. In the presence of a very strong field along the z-axis, the two magnetic moments line up parallel to the z-axis and the potential energy of interaction becomes $U = \left((\mu_i\mu_j/r^3)(1 - 3\cos^2\theta)\right)$.

due to the interaction between identical paramagnetic species in concentrated solids. This interaction is due to the overlap of the orbital wave functions of neighboring electrons. It is, consequently, a short-range interaction and is negligible in dilute samples. For nearest neighbor spins in a concentrated crystal, we can write the interaction energy for isotropic exchange as

$$E = -2J\vec{S}_i \cdot \vec{S}_j \tag{3.61}$$

where J is the exchange constant which will be discussed in more detail in Chapter 6. The result of this interaction is a line that narrows in the center and broadens in the tails giving rise to a line that approximates a Lorentzian instead of a Gaussian [11].

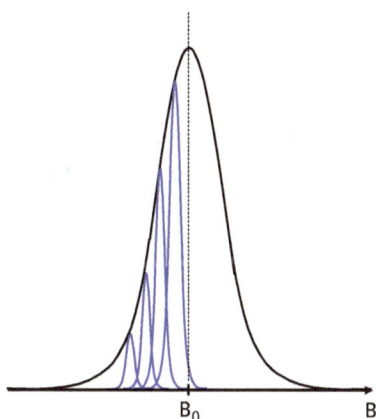

Fig. 3.12: An inhomogeneously broadened line (black line) as an envelope of a large number of "spin packets" (blue line). Each spin packet experiences a slightly different local field leading to small shifts from the resonance condition.

3.6 Saturation effects on the EPR signal

The above relaxation mechanisms must be taken into account when considering the microwave powers that can be used with a particular sample. Care must be taken to ensure that the populations of the two spin states do not get equalized since this leads to saturation and a distortion and decrease of the EPR signal. Saturation depends on the system studied and on the temperature of the sample. For example, systems with organic radicals tend to be easily saturated, whereas systems with transition series ions are not as easy to saturate [12]. The temperature dependence can be seen at cryogenic temperatures where T_1 is slowed down significantly, and energy is not efficiently removed from the spins on the excited state. In such a case, large microwave powers can equalize the populations of the two states causing the rates of stimulated emission and absorption of microwave radiation between the two states to equalize leading to a net zero signal. To preserve the signal, we must allow absorption of microwaves to continue by maintaining a larger spin population on the lower state. As the temperature is lowered, the signal follows a Curie law behavior and the intensity increases as shown in Fig. 3.13. This increase is due to the increasing population difference between the two states as the temperature decreases. This continues until the point where the relaxation times get too long and the populations start to equalize leading to a decrease in signal intensity when saturation starts to dominate. The dependence of the EPR signal on the temperature can be demonstrated by considering the two spin states shown in Fig. 3.2, where we have N total spins populating the two states. If N_- is the number of spins on the $\left|-\frac{1}{2}\right\rangle$ state (lower state) and N_+ the number on the $\left|+\frac{1}{2}\right\rangle$ state (upper state), then the ratio N_-/N_+ at thermal equilibrium in the presence of a magnetic field B is given by the Boltzmann distribution as

$$\frac{N_-}{N_+} = e^{g\beta B/kT} \tag{3.62}$$

The signal intensity is proportional to the difference in populations $\Delta N = N_- - N_+$. Using the total number of spins $N = N_- + N_+$ and dividing the last two expressions by N_+ we get

$$\frac{\Delta N}{N_+} = \frac{N_-}{N_+} - 1 = e^{g\beta B/kT} - 1 \tag{3.63}$$

and

$$\frac{N}{N_+} = \frac{N_-}{N_+} + 1 = e^{g\beta B/kT} + 1 \tag{3.64}$$

Dividing the two expressions term by term yields

$$\frac{\Delta N}{N} = \frac{e^{g\beta B/kT} - 1}{e^{g\beta B/kT} + 1} \tag{3.65}$$

In typical EPR experiments, $g\beta B/kT \ll 1$ allowing us to do a Taylor expansion of the exponential to get

$$e^{g\beta B/kT} \approx 1 + \frac{g\beta B}{kT} \tag{3.66}$$

where only the first two terms in the expansion were retained, leading to

$$\frac{\Delta N}{N} \approx \frac{\frac{g\beta B}{kT}}{2} \Rightarrow \Delta N \approx \frac{Ng\beta B}{2kT} \tag{3.67}$$

which demonstrates the inverse relationship between the signal intensity I and the temperature T since $I \propto \Delta N \propto (1/T)$.

It is obvious from the previous behavior that low microwave powers must be used at low temperatures to avoid signal saturation. The behavior of the EPR derivative signal S with saturation is described by the empirical formula

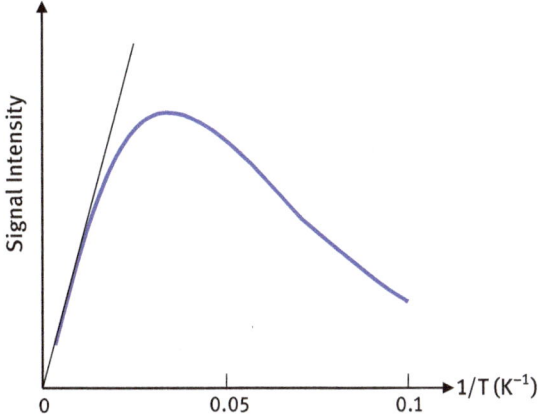

Fig. 3.13: Temperature dependence of the EPR signal. The black straight line indicates the Curie law behavior [14].

$$S = K \frac{\sqrt{P}}{\left(1 + \frac{P}{P_{1/2}}\right)^{b/2}} \tag{3.68}$$

where P is the microwave power, $P_{1/2} = \left(1/\left(g^2 T_1 T_2\right)\right)$ is the microwave power that causes half saturation of the signal, K is a constant and b is the inhomogeneity factor which is equal to 1 for Gaussian lines and 3 for Lorentzian lines [13].

Rearranging the above equation and taking the logarithm of both sides yields

$$\log \frac{S}{P^{1/2}} = -\frac{b}{2}\log\left(P_{1/2} + P\right) + \frac{b}{2}\log\left(P_{1/2}\right) + \log K \tag{3.69}$$

which allows us to plot the log of the signal S to the square root of the power P versus the power P as shown in Fig. 3.14. Under nonsaturating conditions, the curve has a zero slope but as the power is increased and saturation occurs, the curve follows a negative slope going through the point where the intensity drops to half its nonsaturating value $P_{1/2}$. The slope of this line depends on which relaxation mechanism dominates: the slope is more negative for a Gaussian curve than for a Lorentzian curve but both intersect the horizontal line at the same $P_{1/2}$ value [13].

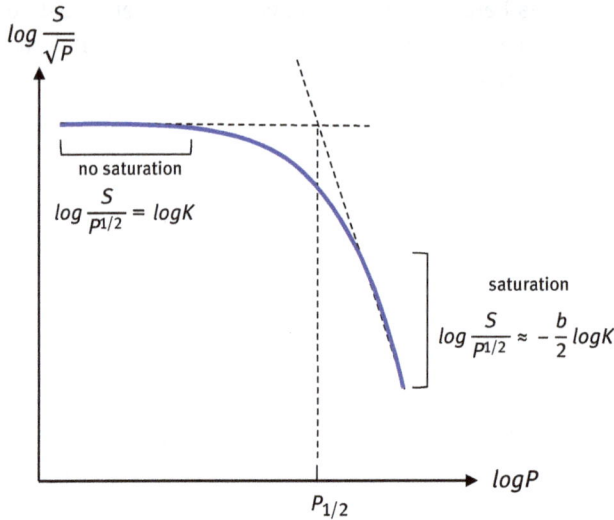

Fig. 3.14: Microwave power saturation curve of an EPR line at constant temperature. The horizontal part of the curve indicates the region of no saturation which occurs when $P \ll P_{1/2}$ where $P_{1/2}$ is the half-saturation point [12].

It is clear from the above discussion that EPR experiments should be carried out far from saturating conditions.

3.7 Transition probabilities

The absorption of energy for transitions between quantum states $|\psi_m\rangle(=|m\rangle)$ and $|\psi_n\rangle(=|n\rangle)$, which are eigenfunctions of the static Hamiltonian $\mathcal{H}_0$, depends on the following factors: (i) the energy density of the microwave radiation, (ii) the population of the states and (iii) the square of the transition matrix element [9]. In the phenomenological discussion of EPR in Chapter 1, we saw that according to classical arguments, the EPR phenomenon occurs when the microwave magnetic field $\vec{B}_1$ is orthogonal to the external magnetic field. To determine the transition probability for transitions between states $|n\rangle \rightarrow |m\rangle$ $(E_m > E_n)$ induced by an oscillating (time-dependent) microwave magnetic field in a quantum mechanical framework, we must use the time-dependent Schrödinger equation

$$\mathcal{H}|\Psi\rangle = i\hbar \frac{\partial}{\partial t}|\Psi\rangle \tag{3.70}$$

where the spin Hamiltonian is now given by $\mathcal{H} = \mathcal{H}_0 + \mathcal{H}_1$ where in addition to the time-independent interactions such as the Zeeman and zero-field splitting terms represented by the term $\mathcal{H}_0$, now includes a time-dependent term $\mathcal{H}_1$ to account for the interaction between the oscillating magnetic field component $\vec{B}_1$ of the microwave radiation and the spin magnetic moment

$$\mathcal{H}_1 = \beta \vec{B}_1 \cdot g \cdot \vec{S} \tag{3.71}$$

where, in general, the g-value is expressed by the tensor g which will be discussed in the next section. The transition probability between the two states is then given by

$$W_{nm} \propto N(n)\rho(v)g(v)\left|\langle m|\mathcal{H}_1|n\rangle\right|^2 = \beta^2 N(n)\rho(v)g(v)\left|\langle m|\vec{B}_1 \cdot g \cdot \vec{S}|n\rangle\right|^2 \tag{3.72}$$

where $\rho(v)$ is the energy density of the microwave radiation, which is proportional to v^2 in air and is proportional to B_1^2 within the microwave cavity. $N(n)$ is the number of electrons on the $|n\rangle$ state and $g(v)$ is the normalized line shape function, which can either correspond to a Lorentzian $g_L(v)$ or Gaussian $g_G(v)$ line shape and is given by the expressions

$$g_L(v) = \frac{2T_2}{1 + T_2^2 4\pi^2(v - v_0)^2} \tag{3.73}$$

$$g_G(v) = 2T_2 e^{-4\pi T_2^2 (v - v_0)^2} \tag{3.74}$$

where T_2 is the transverse (spin–spin) relaxation time [2]. For a system of N spins in thermal equilibrium between the two states, the maximum power absorbed per unit volume during resonance at frequency v_0 is proportional to

$$P \propto \Delta N\rho(v)g(v)hv_0\left|\langle m|\vec{B}_1 \cdot g \cdot \vec{S}|n\rangle\right|^2 \tag{3.75}$$

where $\Delta N = N(n) - N(m)$ is the difference in the spin populations of the two states which reduces to $\Delta N \approx Nhv_0/2k_BT$ when $hv_0 \ll k_BT$. This formula indicates that the intensity of the EPR line is proportional to the number of spins N, the frequency of the incident radiation v_0, the square of the magnetic field component of the microwave field B_1^2 and the matrix element $\langle m|\vec{B}_1 \cdot g \cdot \vec{S}|n\rangle$. It is inversely proportional to temperature T, which means that the signal will be stronger at high spin densities, high microwave frequencies and power (larger $\rho(v)$ in cavity), and low temperatures. Of course, the microwave power cannot be increased indefinitely in order to make sure that thermal equilibrium is maintained between the two states so as to avoid saturation effect as discussed in the previous section. The effect of the matrix element $\langle m|\vec{B}_1 \cdot g \cdot \vec{S}|n\rangle$ on the signal will be discussed below for the case of both Kramers and non-Kramers ions in cubic and axial symmetries (these symmetries are defined in Section 3.11). An $S = 1/2$ spin will be considered as a representative Kramers system and an $S = 1$ spin for a non-Kramers system but the discussion can be extended to more general systems.

3.7.1 $S = 1/2$ system in cubic symmetry

This system provides a straightforward application of the above discussion where the transitions are between the states $|-\frac{1}{2}\rangle$ and $|+\frac{1}{2}\rangle$. In this case, the g-tensor is isotropic allowing us to write the time-dependent Hamiltonian as

$$\mathcal{H}_1 = \beta g \vec{B}_1 \cdot \vec{S} \tag{3.76}$$

With the static external field $\vec{B}_0$ applied along the z-axis we get $\vec{B}_1 \cdot \vec{S} = B_{1z} S_z$, yielding matrix elements of the form $\langle m|\vec{B}_1 \cdot \vec{S}|n\rangle = \langle m|B_{1z} S_z|n\rangle = B_{1z}\langle m|S_z|n\rangle$. We can now consider the matrix elements that result from two different orientations of $\vec{B}_1$ relative to the coordinate axes.

$$\vec{B}_1 \parallel z - \text{axis}$$

The matrix elements have the form $\langle +\frac{1}{2}|S_z|-\frac{1}{2}\rangle$. S_z operating on $|-\frac{1}{2}\rangle$ yields

$$S_z\left|-\frac{1}{2}\right\rangle = -\frac{1}{2}\left|-\frac{1}{2}\right\rangle \tag{3.77}$$

resulting in

$$\left\langle +\frac{1}{2}\middle|S_z\middle|-\frac{1}{2}\right\rangle = -\frac{1}{2}\left\langle +\frac{1}{2}\middle|-\frac{1}{2}\right\rangle = 0 \tag{3.78}$$

since $|-\frac{1}{2}\rangle$ and $|+\frac{1}{2}\rangle$ are orthogonal eigenfunctions. It is obvious from this that when the microwave magnetic field is parallel to the external field $\vec{B}_0$, the transition probability is zero.

$$\vec{B}_1 \parallel x - \text{axis}$$

The matrix elements now have the form $\langle +\frac{1}{2}|S_x|-\frac{1}{2}\rangle$. The spin operator S_x can be expressed as

$$S_x = \frac{1}{2}(S_+ + S_-) \tag{3.79}$$

which when operating on the $|-\frac{1}{2}\rangle$ state yields

$$S_x\left|-\frac{1}{2}\right\rangle = \frac{1}{2}(S_+ + S_-)\left|-\frac{1}{2}\right\rangle = \frac{1}{2}\left|+\frac{1}{2}\right\rangle \tag{3.80}$$

The resulting matrix element then is

$$\left\langle +\frac{1}{2}\middle|S_x\middle|-\frac{1}{2}\right\rangle = \frac{1}{2}\left\langle +\frac{1}{2}\middle|+\frac{1}{2}\right\rangle = \frac{1}{2} \tag{3.81}$$

which means that the transition probability is now not equal to zero. A similar result is obtained when $\vec{B}_1 \parallel y - $ axis as well leading to the conclusion that transitions occur when $\vec{B}_1 \perp \vec{B}_0$.

3.7.2 $S = 1/2$ system in axial symmetry

The above discussion can be extended to the case of an effective spin $1/2$ system in an axial symmetry with $\vec{H}_0$ along the z-axis which is also the symmetry axis. The matrix elements are then given by [15]

$$B_1 g_{\parallel} \left\langle +\frac{1}{2} \middle| S_z \middle| -\frac{1}{2} \right\rangle = 0 \tag{3.82}$$

for $\vec{B}_1 \parallel z - $ axis

$$B_1 g_{\perp} \left\langle +\frac{1}{2} \middle| S_x \middle| -\frac{1}{2} \right\rangle = \frac{1}{2} H_1 g_{\perp} \tag{3.83}$$

for $\vec{B}_1 \parallel x - $ axis

$$B_1 g_{\perp} \left\langle +\frac{1}{2} \middle| S_y \middle| -\frac{1}{2} \right\rangle = -\frac{i}{2} H_1 g_{\perp} \tag{3.84}$$

for $\vec{B}_1 \parallel y - $ axis, demonstrating that in this case transitions again occur when $\vec{B}_1 \perp \vec{B}_0$. It is easy to show by following similar arguments that the same result can be obtained with $S_{\text{eff}} = 1/2$ system in rhombic symmetry.

The above results demonstrate that the classical conclusion that $\vec{B}_1$ must be perpendicular to $\vec{B}_0$ for EPR transitions to occur also has a quantum mechanical foundation. This result can be generalized to all Kramers systems where the requirement for EPR transitions satisfying the selection rule $\Delta m_S = \pm 1$ is to orient the external magnetic field relative to the microwave cavity in such a way as to have $\vec{B}_1 \perp \vec{B}_0$.

3.7.3 $S = 1$ system

Non-Kramers ions (ions with integer spin) have a different behavior than Kramers ions and yield EPR spectra that, in many cases, are highly anisotropic. Transitions that are forbidden in Kramers ions are allowed in these ions under certain conditions. A system with an effective spin $S = 1$ will be used as an example here. In cubic symmetry, the crystal field splits the $S = 1$ manifold into the pure states $|0\rangle$ and $|\pm 1\rangle$ when $\vec{B}_0$ is oriented along the z-axis. Transitions obeying the selection rule $\Delta m_S = \pm 1$ are allowed between the states $|-1\rangle \leftrightarrow |0\rangle$ and $|+1\rangle \leftrightarrow |0\rangle$ with $\vec{B}_1 \perp \vec{B}_0$, and no transitions between states $|-1\rangle \leftrightarrow |+1\rangle$ can occur for either orientation of $\vec{B}_1$ relative to $\vec{B}_0$.

E

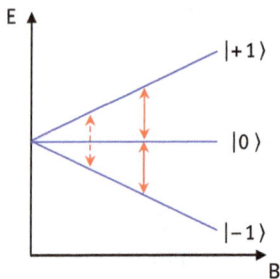

Fig. 3.15: Splitting of the $S = 1$ manifold inside a crystal field of cubic symmetry. The states are pure $|0\rangle$ and $|\pm 1\rangle$ states, and there is no zero-field splitting. Allowed transitions are shown by solid double arrows. The dashed double arrow indicates forbidden transitions.

In an axial crystal field, the states with $\vec{B}_0$ along the z-axis remain pure but there is now zero-field splitting that separates the states $|0\rangle$ and $|\pm 1\rangle$ states. The same transitions occur as in the cubic case. When $\vec{B}_0$ is perpendicular to the z-axis, the states are now admixed and are given by

$$|\psi_0\rangle = \frac{1}{\sqrt{2}}(|+1\rangle - |-1\rangle) \tag{3.85}$$

$$|\psi_+\rangle = a(|+1\rangle - |-1\rangle) + b|0\rangle \tag{3.86}$$

$$|\psi_-\rangle = a'(|+1\rangle - |-1\rangle) - b'|0\rangle \tag{3.87}$$

where a, b, a' and b' are positive constants that depend on $g\beta B$ and D. The mixing of the states allows matrix elements of the form $\langle \psi_+|S_z|\psi_0\rangle$ and $\langle \psi_-|S_z|\psi_0\rangle$ to have nonzero values yielding nonzero probability densities for transitions $\psi_+ \leftrightarrow \psi_0$ and $\psi_- \leftrightarrow \psi_0$ when $\vec{B}_1 \perp \vec{B}_0$. When $\vec{B}_1 \parallel \vec{B}_0$, normally forbidden transitions between the states $\psi_- \leftrightarrow \psi_+$ are allowed since matrix elements of the form $\langle \psi_-|S_x|\psi_+\rangle$ are non-zero, yielding finite probability densities for such transitions. These transitions happen at low applied fields as shown in Fig. 3.15 since at high fields the applied field determines the axis of quantization resulting in linearly divergent energy levels and making such transitions forbidden. This is called parallel-mode EPR when $\vec{B}_1 \parallel \vec{B}_0$ results in typically highly anisotropic EPR lines with large g-values. These will be further explored in the discussion on integer spin systems in Chapter 6.

3.8 The g-value

The g-value was first introduced in terms of the magnetic moment of the electron in Chapter 2. It is one of the experimental quantities that can be determined from the EPR spectrum using the resonance condition $h\nu = g\beta B$. The g-value provides important information about the behavior of unpaired electrons in a sample. We have already seen that the g-value for a free (unbound) electron is equal to $g_e = 2.0023$, but for a bound electron the g-value can deviate significantly from this value. The reason for this is the fact that for a bound electron we need to take into account the effects of the orbital motion of the electron. The orbital angular momentum of the bound electron interacts with both the spin angular momentum of the electron through spin–orbit coupling as well as the surrounding ligands of the atom, giving rise to a shift of g relative to g_e along with anisotropy in g.

3.8.1 Deviations of g relative to free electron value

Spin–orbit coupling leads to deviations of g-values from the free electron value of $g_e = 2.0023$. As discussed in Chapter 2, this is the result of an interaction between the orbital and spin magnetic moments of bound electrons. Even ground states where the orbital angular momentum is zero (due to the "quenching" of the orbital angular momentum caused by the crystal field), where the g-value should be expected to be equal to g_e, the spin–orbit coupling introduces small admixtures from the excited states that result in a small nonzero orbital angular momentum introduced into the ground state [16]. Consequently, spin–orbit coupling makes the g-value a characteristic property of the particular ion, the oxidation state of the ion and the crystal field experienced by the ion that produces the quenching of the orbital contribution [17]. The value of g also depends on the orientation of the magnetic field relative to the crystal field symmetry axes, introducing anisotropy into g. Since EPR provides a direct measurement of g, it makes it the perfect technique to provide information on the properties of the ion.

In a Russell–Saunders scheme when spin–orbit coupling is present as a perturbation, second-order perturbation theory shows that the g-tensor in the presence of spin–orbit coupling is shifted relative to g_e according to [18]

$$\boldsymbol{g} = g_e \boldsymbol{I} + 2\lambda \boldsymbol{\Lambda} \tag{3.88}$$

where λ is the spin–orbit coupling constant, $\boldsymbol{I}$ is the identity matrix and the second-rank tensor $\boldsymbol{\Lambda}$ is given by

$$\boldsymbol{\Lambda} = \sum_n \frac{\langle 0|\hat{L}|n\rangle \langle n|\hat{L}|0\rangle}{E_0 - E_n} \tag{3.89}$$

$|0\rangle$ represents the ground state of the system with energy E_0 and $|n\rangle$ represents an excited state with energy E_n. $\hat{L}$ is the orbital angular momentum operator. This expression is applied when the ground state is an orbital singlet. The tensor $\boldsymbol{\Lambda}$ introduces orbital degeneracy into the ground state that causes the deviations from g_e. In other words, spin–orbit coupling connects the ground state with excited states, resulting in small admixtures of the excited state wave function introduced into the ground state that slightly change its magnetic properties. The tensor $\boldsymbol{\Lambda}$ can be expressed as a 3×3 matrix with elements Λ_{ij}:

$$\Lambda_{ij} = \sum_n \frac{\langle 0|\hat{L}_i|n\rangle \langle n|\hat{L}_j|0\rangle}{E_0 - E_n} \tag{3.90}$$

with $\hat{L}_i$ and $\hat{L}_j$ the orbital angular momentum operators correspond to the particular coordinate i or $j = x$, y, z. These matrix elements are nonzero when they connect states that are related through rotations about the coordinate axes x, y and z. For example, rotations of the d_{xy} orbital about x, y, z result, within a numerical factor, in the orbitals d_{xz}, d_{yz}, $d_{x^2-y^2}$, respectively [19]. So, the only nonzero matrix elements of Λ when an electron is on the d_{xy} ground orbital will be the ones between d_{xy} and the d_{xz}, d_{yz}, $d_{x^2-y^2}$ orbitals. This allows us to express eq. (3.88) for g in a principal axis system (discussed in Section 3.8.2) in terms of the numerical values of the matrix elements represented by an integer n and the energy separation Δ between the ground orbital and the appropriate excited orbital as

$$g_i = g_e \pm n\frac{\lambda}{\Delta} \tag{3.91}$$

where $i = x$, y, z. The plus or minus sign is used depending on whether the shell is less or more than half-full, respectively. The numerical values of n can be easily obtained from a useful construct called the *magic pentagon* as illustrated in Fig. 3.16. As an example, consider a d^1 system with the electron on the ground orbital d_{xy}. The energy-level diagram of the orbital states could be represented in Fig. 3.17. According to the magic pentagon, d_{xy} is connected to the d_{xz}, d_{yz}, $d_{x^2-y^2}$ orbitals with the values of n given by 2, 2 and 8, respectively. The g-values for this configuration are then given by

$$g_x = g_e - 2\frac{\lambda}{\Delta_1}, \quad g_y = g_e - 2\frac{\lambda}{\Delta_2}, \quad g_z = g_e - 8\frac{\lambda}{\Delta_3} \tag{3.92}$$

where Δ_1, Δ_2, Δ_3 are the energy separations between d_{xy} and the d_{xz}, d_{yz}, $d_{x^2-y^2}$ orbitals, respectively, as shown in Fig. 3.17.

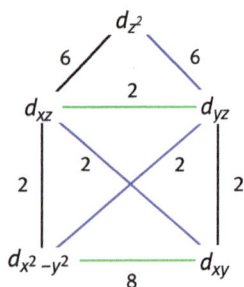

Fig. 3.16: The magic pentagon with d-orbitals at its apexes. The lines connect orbitals that are rotationally related under rotations about the x –, y –, z – axes. The number on each line corresponds to the value of the appropriate matrix element Λ_{ij}. The rotations about the coordinate axes are labeled as follows: blue = x-axis, black = y-axis and green = z-axis.

In the case of an axial geometry where d_{xz} and d_{yz} are degenerate, then $\Delta_1 = \Delta_2 = \Delta$ making

$$g_x = g_y = g_\perp = g_e - 2\lambda/\Delta \tag{3.93}$$

and

$$g_z = g_\parallel = g_e - 8\lambda/\Delta \tag{3.94}$$

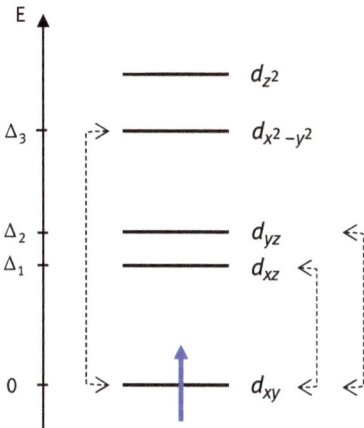

Fig. 3.17: Energy-level diagram for the orbitals of a d^1 ion inside a crystal field that completely removes the orbital degeneracy. The electron resides on the singlet ground orbital d_{xy} which is connected to the d_{xz}, d_{yz}, $d_{x^2-y^2}$ orbitals via rotations about the $x-$, $y-$, $z-$ axes. These excited states contribute to the orbital magnetism of the ground singlet that leads to a shift of the $g-$ values of the ground state away from g_e.

3.8.2 *g*-Value anisotropy

In general, the *g*-value is derived from a tensor with nine components that reflects the anisotropy of the molecule or crystal. It is represented by a symmetric tensor of rank 2 and can be expressed as the 3 × 3 matrix

$$\mathbf{g} = \begin{pmatrix} g_{xx} & g_{xy} & g_{xz} \\ g_{yx} & g_{yy} & g_{yz} \\ g_{zx} & g_{zy} & g_{zz} \end{pmatrix} \tag{3.95}$$

For this general case, the spin magnetic moment can then be written in terms of its components as follows:

$$\begin{pmatrix} \mu_x \\ \mu_y \\ \mu_z \end{pmatrix} = \beta \begin{pmatrix} g_{xx} & g_{xy} & g_{xz} \\ g_{yx} & g_{yy} & g_{yz} \\ g_{zx} & g_{zy} & g_{zz} \end{pmatrix} \begin{pmatrix} S_x \\ S_y \\ S_z \end{pmatrix} \tag{3.96}$$

Any tensor of rank 2 can be reduced to a diagonal form with a proper choice of coordinate axes called the principal axes. The principal axes can be chosen to coincide with the (x, y, z) coordinates to yield the diagonal tensor

$$\boldsymbol{g} = \begin{pmatrix} g_x & & \\ & g_y & \\ & & g_z \end{pmatrix} \tag{3.97}$$

where $g_x = g_{xx}$, $g_y = g_{yy}$, $g_z = g_{zz}$. When $g_x = g_y = g_z$, the g-tensor is isotropic and independent of the choice of axes. When $g_x = g_y \neq g_z$, the g-tensor is axially symmetric and when $g_x \neq g_y \neq g_z$ the g-tensor represents lower symmetries (rhombic). When $\boldsymbol{g}$ is in a general direction relative to the principal axes, it is given by

$$g = \left(l^2 g_x^2 + m^2 g_y^2 + n^2 g_z^2 \right)^{1/2} \tag{3.98}$$

where (l, m, n) are the direction cosines relative to the principal axes.[3] In spherical coordinates, using the angle definitions in Chapter 2, we can write the above expression as follows [8]:

$$g = \left(g_x^2 \sin^2\theta \cos^2\phi + g_y^2 \sin^2\theta \sin^2\phi + g_z^2 \cos^2\theta \right)^{1/2} \tag{3.99}$$

For an effective spin $S = 1/2$ system, we can use the above discussion for the g-value in the Zeeman Hamiltonian, which can be written as

$$\mathcal{H} = \beta \vec{S} \cdot \boldsymbol{g} \cdot \vec{B} \tag{3.100}$$

which in a principal axes system reduces to

$$\mathcal{H} = \beta \left(g_x S_x B_x + g_y S_y B_y + g_z S_z B_z \right) \tag{3.101}$$

For an arbitrary direction of $\vec{B}$ given by the direction cosines (l, m, n), the energy levels are equal to

$$E = \pm \frac{1}{2} g \beta B \tag{3.102}$$

where g is given by eq. (3.99). For the special case of axial symmetry with the z-axis as the symmetry axis with $\vec{B}$ making an angle θ with the z-axis, as shown in Fig. 3.18, the g-value becomes [20]

$$g = \left(g_z^2 \cos^2\theta + g_\perp^2 \sin^2\theta \right)^{1/2} \tag{3.103}$$

where $g_\perp = g_x = g_y$.

---

3 The direction cosines (l_x, l_y, l_z) of a general three-dimensional vector with respect to the coordinate axes (x, y, z) are defined as $l_i^2 = \cos^2\theta_i$, where $i = x$, y, z and θ_i is the angle between g and the coordinate axis i. From this definition, it follows that $l_x^2 + l_y^2 + l_z^2 = 1$. The notation used in this book assumes $l_x = l$, $l_y = m$, $l_z = n$.

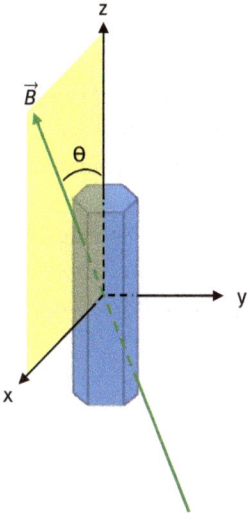

Fig. 3.18: Orientation of the magnetic field $\vec{B}$ relative to the principal axes of an axial crystal with an axis of symmetry along the *z*-axis. The field rotates in the *xz*-plane making an angle θ with the *z*-axis. When $\theta = 0° \Rightarrow \vec{B} \parallel z$ the EPR spectrum will yield the value of $g_\parallel$, and when $\theta = 90° \Rightarrow \vec{B} \perp z$ the EPR spectrum will yield the value of $g_\perp$.

The previous discussion demonstrates that the g-value can be anisotropic based on how its three components are related. This anisotropy in *g* is introduced by the molecular symmetry (arrangement of nuclei relative to the electron orbital of the paramagnetic ion) and gives rise to anisotropy in the EPR line. In addition, for transition series and rare earth ions, the ligands contribute to the anisotropy. The experimental EPR line shown in Fig. 3.19(a) is an example of an isotropic spectrum of a crystalline sample, displayed as the first derivative of the absorption signal. This occurs when the paramagnetic ion experiences cubic symmetry where all directions in space are equivalent, and the g-value is isotropic, that is, $g_x = g_y = g_z = g$. In this case, the g-value can be calculated from the field position B_0 of the EPR line using the resonance condition

$$g = \frac{h\nu}{\beta B_0} \tag{3.104}$$

If the symmetry is axial, then one axis of the molecule is unique (typically the *z*-axis) while the other two orientations are equivalent producing two unique g-values: $g_z = g_\parallel$ and $g_x = g_y = g_\perp$. In this case, the field position of the EPR line will depend on the orientation of the external magnetic field relative to the symmetry axes of the crystal. If we assume that $g_\parallel < g_\perp$, then according to the resonance condition, the EPR line when the magnetic field is parallel to the *z*-axis will occur at a lower field position than when the magnetic field is oriented in the *xy*-plane. This is demonstrated in Fig. 3.19(b), where the EPR line shifts from magnetic field position $B_\parallel$ to $B_\perp$ as the magnetic field is rotated away from the *z*-axis and toward the *xy*-plane. The g-value for a general orientation of the magnetic field is determined from

$$g = \left(g_\parallel^2 \cos^2\theta + g_\perp^2 \sin^2\theta\right)^{1/2} = \frac{h\nu}{\beta B} \tag{3.105}$$

where θ is the angle between the magnetic field and the crystalline axis of symmetry. To determine the $g_\parallel$ and $g_\perp$ values, we record the spectrum with $\vec{B}$ parallel ($\theta = 0°$) and perpendicular ($\theta = 90°$) to the z-axis, respectively. If the symmetry is rhombic, there are no equivalent axes, and three different g-values $g_x \neq g_y \neq g_z$ are obtained. In this case, the EPR line will shift from B_z to B_y and B_x as the magnetic field is rotated away from the z-axis of the crystal and toward the y- and x-axes as demonstrated in Fig. 3.19(c). The g-value now has the more general form given in

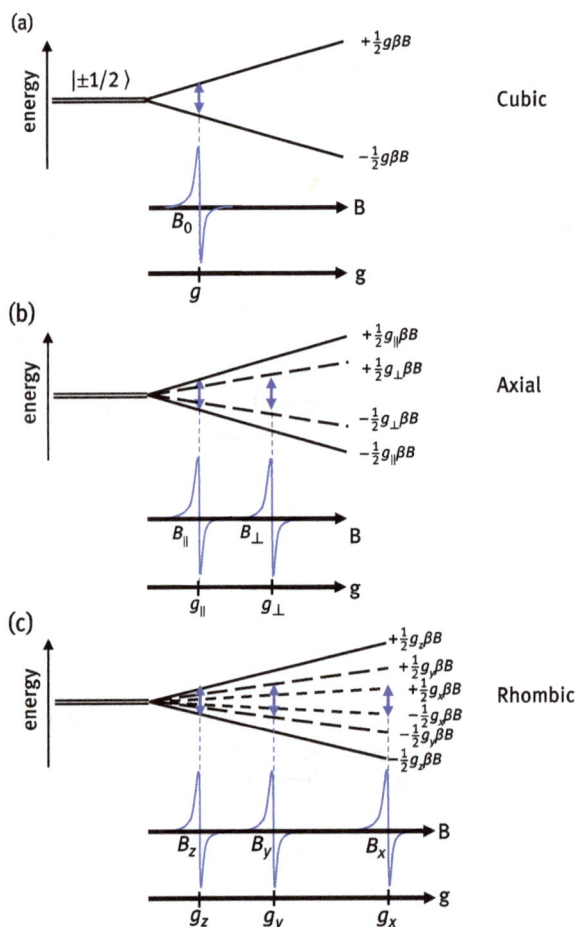

Fig. 3.19: Energy-level splittings of an $S = 1/2$ ion in three different symmetries: (a) isotropic, (b) axial and (c) rhombic. The lines below the energy-level diagrams show the expected behavior of the EPR spectrum in the three symmetries.

the previous section. In order to experimentally determine the three g-values, we need to record EPR spectra with $\vec{B}$ parallel to each of the coordinate axes.

Example: An example of the angular variation of EPR lines is provided by Ce(III) doped in the diamagnetic host lanthanum trifluoromethanesulfonate (LaTFMS). Crystals of this compound have an axial C_{3h} symmetry with small C_{3v} distortions that allow EPR transitions to occur within the two lowest doublets which are separated by $\Delta/k_B = 9.6$ K. Spin–orbit coupling splits the 2 F level of Ce(III) into two energy manifolds $J = 7/2$ and $J = 5/2$ separated by ~2,000 cm^{-1}. Since the two manifolds are so widely separated in energy, we can only consider the lower manifold in this discussion. The $J = 5/2$ energy manifold is split into three Kramers doublets by the axial crystal field as shown in Fig 3.20(a). Lattice distortions introduce admixtures between the two lowest doublets that allow transitions to occur within each of these doublets. Due to strong spin–lattice coupling, the EPR lines of rare earth ions broaden substantially at higher temperatures; consequently, signals can only be observed at very low temperatures. At X-band frequencies and liquid helium temperatures, two EPR lines are observed when the magnetic field is parallel to the crystalline c-axis at 1.727 and 6.738 kG resulting from transitions within the two lowest doublets which can be treated as effective spin 1/2 states. As the magnetic field is rotated away from the c-axis and toward the perpendicular ab-plane, the line positions shift in field as shown in Fig 3.20(b). A fit of the data for each line to eq. (3.105) yields the g-values $g_\parallel = 3.813$ and $g_\perp = 0.262$ for the ground doublet and $g_\parallel = 0.977$, $g_\perp = 2.151$ for the first excited state [21].

In general, an EPR signal from a $S = \frac{1}{2}$ system can show one of four different types of polycrystalline patterns (also referred to as a powder pattern) displayed in Fig. 3.21. Most commonly, a measurement is recorded on a solution, frozen solution, or powder sample. In such sample, all orientations of the molecule with respect to the magnetic field are possible. The polycrystalline pattern depends on the g-values and shows characteristic features at magnetic fields corresponding to the principal g-values as shown in Fig. 3.21. For metal complexes in low symmetry, the orientation of the principal axis system of the g-tensor with respect to the spatial arrangement of atoms will not be known, but can be determined from measurement of single crystals. The shape of these signals will be further discussed in Section 3.11 where symmetry effects on the EPR spectrum are explored.

3.8.3 g-Strain broadening

The EPR spectra of many metal-containing biomolecular systems with anisotropic g-values exhibit inhomogeneous broadening effects that are explained by the concept of "g-strain" [23]. This refers to strains within the biomolecule that result in small changes in the bond lengths and angles of the ligands that form the metal environment. These changes can produce small shifts in the resonance magnetic

(a)

(b)

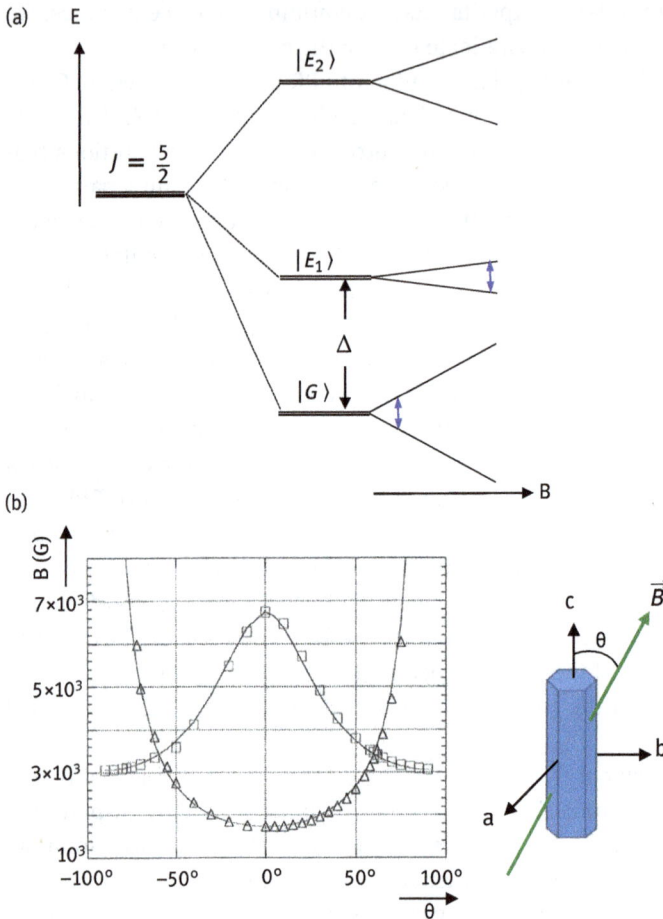

Fig. 3.20: (a) Energy-level diagram for the $J = 5/2$ manifold of Ce(III). Transitions occur within the two lowest doublets (indicated by the blue arrows) which can be treated as states with an effective spin of 1/2. These transitions give rise to the two observed EPR lines at $T = 4.2$ K. No transitions are allowed in the second excited doublet. (b) Angular variation of the EPR lines in Ce(III) doped in LaTFMS recorded at X-band. Triangles correspond to the field positions of the ground state line and squares for the first excited state line. The solid curves are fits of the data to eq. (3.105). The orientation of the external magnetic field B relative to the crystalline axes is also shown. The crystals have a hexagonal habit. The z-axis is defined along the c-axis of the crystal.

field of a particular resonance line. This results in a distribution of g-values, usually Gaussian in nature that produces a broadening of the EPR line. In a rhombic environment, we can have small distributions of g-values centered at each of the values g_x, g_y and g_z. The broadening of each line is anisotropic since a small increase in g results in a small change in the resonance field B_0 when B_0 is small and a large increase when B_0 is large. The result of this on an EPR spectrum such as the one in

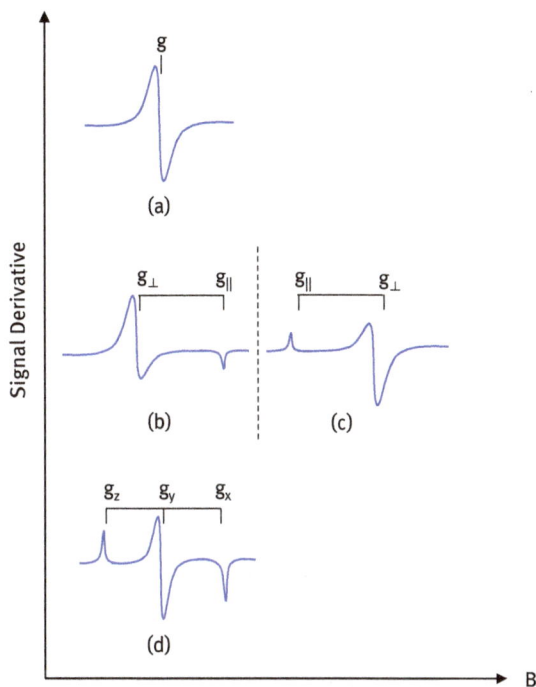

Fig. 3.21: Typical EPR powder spectra for various molecular symmetries: (a) isotropic, (b) axial with $g_x = g_y > g_z$, (c) axial with $g_x = g_y < g_z$, and (d) rhombic. The geometric shapes corresponding to these symmetries are: (a) sphere, (b) oblate spheroid, (c) prolate spheroid and (d) distorted ellipsoid [22].

Fig. 3.21(d) would be to broaden the line at g_x much more than the line at g_z. In some cases, the line at g_x is so broad that it is barely visible relative to the baseline. An example of this is shown in Fig. 3.22 for the EPR spectrum of horse heart cytochrome c where a spread of g-values of ~0.2 produces a ~140 G linewidth for g_z and ~900 G for g_x [24].

3.9 Zero-field splittings

For paramagnetic ions with $S > 1/2$, the degeneracy of the spin energy states is removed under the influence of the ligand field experienced by the ion. This occurs in the absence of an external magnetic field and in ligand fields of symmetry lower than octahedral. This "zero-field splitting" can produce energy-level separations ranging from a few wave numbers to many tens of wave numbers and, consequently, determine the characteristics of the EPR spectra of paramagnetic ions in solid systems. In half-integer spin systems (Kramers ions), the level degeneracy is

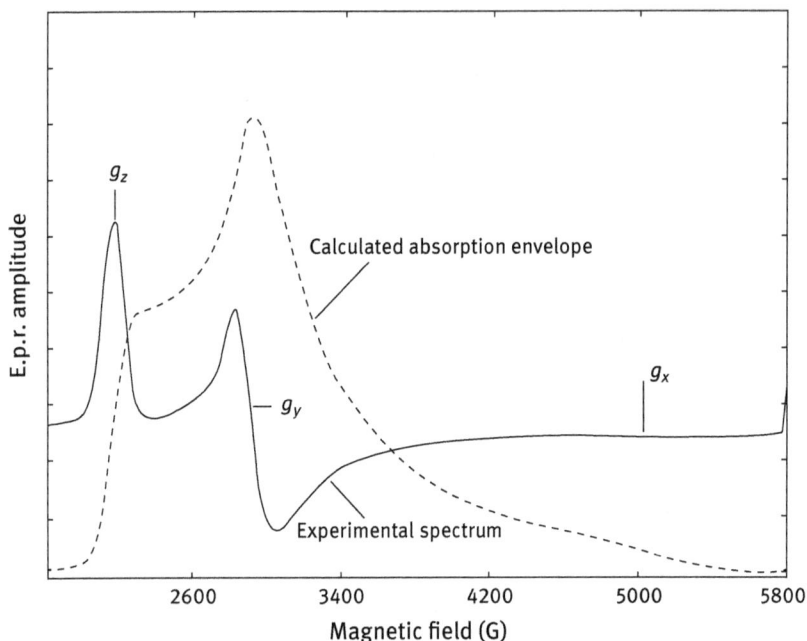

Fig. 3.22: EPR spectrum of horse heart cytochrome c recorded at a frequency of 9.24 GHz at a temperature of 10 K. The extremely small signal amplitude associated with the high field line at g_x is a consequence of g-strain in this molecule.

removed down to doublets (Kramers doublets), and in integer spin systems (non-Kramers ions), the degeneracy can be completely removed down to singlet levels. In this section, we will discuss the basic effects that this kind of splitting has on the EPR spectrum. The zero-field Hamiltonian is written as

$$\mathcal{H} = \vec{S} \cdot \boldsymbol{D} \cdot \vec{S} \tag{3.106}$$

where **D** is a symmetric traceless tensor representing the crystal field. This Hamiltonian is a result of spin–orbit coupling and is sensitive to the symmetry of the ligand environment. Adding the Zeeman term yields the complete Hamiltonian [25]

$$\mathcal{H} = \vec{S} \cdot \boldsymbol{D} \cdot \vec{S} + \beta \vec{S} \cdot \boldsymbol{g} \cdot \vec{B} \tag{3.107}$$

where **g** is the g-tensor. Since **D** is a traceless tensor it can be separated into two independent terms D and E, the axial and rhombic zero-field parameters respectively. We can express these parameters in terms of the principal values of the **D** tensor as $D = 3D_{zz}/2$ and $E = |D_{xx} - D_{yy}|/2$. For a magnetic field along the z-axis we can then write the Hamiltonian as

$$\mathcal{H} = g_z \beta S_z B_z + D\left[S_z^2 - \frac{S(S+1)}{3}\right] + E\left(S_x^2 - S_y^2\right) \qquad (3.108)$$

For axial symmetry, $E = 0$ and this Hamiltonian will remove the spin degeneracy of a Kramers system down to the level of doublets, as mentioned above.

3.9.1 Half-integer spin (Kramers) systems

The $S = 3/2$ and $S = 5/2$ spin systems which appear in the majority of experimental situations will be discussed below. The discussion can easily be extended to the less prevalent $S = 7/2$ and $S = 9/2$ systems. A word on rhombograms and their utility in determining the rhombicity of a particular system and assigning signals to specific doublets is also included at the end of this section.

3.9.1.1 $S = 3/2$

The simplest example of a half-integer spin system in this category occurs when $S = 3/2$. An application of this is found when a Cr(III) ion with a d^3 configuration is placed in an axially distorted octahedral environment. The diagonalization of the zero-field Hamiltonian in the $|m_S\rangle$ basis is done in Chapter 6 and yields the energy values

$$E_{\pm 1/2} = -D \pm g_z \beta B_z/2 \qquad (3.109)$$

$$E_{\pm 3/2} = D \pm 3g_z \beta B_z/2 \qquad (3.110)$$

We see that a crystal field with axial symmetry partially removes the fourfold degeneracy by separating the $|\pm 1/2\rangle$ and the $|\pm 3/2\rangle$ states by an energy equal to 2D. This is illustrated in Fig. 3.23 for small and large values of D where the $|\pm 1/2\rangle$ state was taken as the origin of energy values. The value and sign of D depends on the nature, arrangement and distance of the ligands relative to the paramagnetic ion. The diagrams drawn are for $D > 0$. For $D < 0$, the $|\pm 3/2\rangle$ state is lower in energy [26]. It should be noted that zero-field splitting introduces paramagnetic anisotropy resulting in shifts in the EPR spectra as the magnetic field is rotated relative to the principal axes of the molecule. In addition, when the magnetic field is rotated away from the z-axis, which defines the axial symmetry, there is a competition between the magnetic and crystal fields in which one determines the axis of quantization. When the magnetic field is stronger than the crystal field ($g\beta B \gg 2D$), the magnetic field dominates and the energy levels are linearly split in a manner similar to the one shown in Fig. 3.23. If, however, the magnetic and crystal fields are of comparable strength ($g\beta B \approx 2D$), then the energy levels are curved and can lead to significant shifts in the EPR lines. Another artifact is the so-called mixing of states that occurs in this case where elements of one doublet are admixed within the other, allowing forbidden transitions to occur as seen earlier.

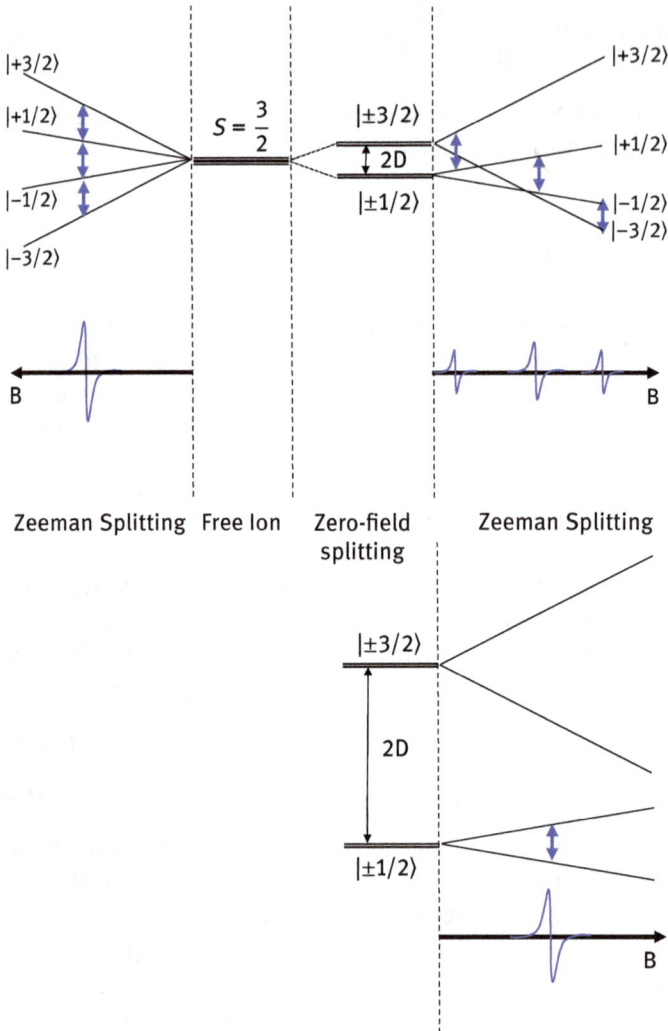

Fig. 3.23: Zero-field splitting for an $S = 3/2$ ion. The diagram to the left of the free ion shows the Zeeman splitting in the absence of zero-field splitting which results in three transitions according to the selection rule $\Delta m_s = \pm 1$ occurring at the same field value (blue arrows) that yields overlapping EPR lines. In the presence of a small zero-field splitting (zfs), the energy manifold splits into two Kramers doublets resulting in inter- and intradoublet transitions that do not overlap in magnetic field position. The result is three EPR lines. For large zfs, the $|\pm 3/2\rangle$ state can be quite removed from the $|\pm 1/2\rangle$ resulting in only one transition and one EPR line in the spectrum. The line in this case can be quite intense at low temperatures when $D \gg k_B T$.

3.9.1.2 $S = 5/2$

A further example of zero-field splitting is given for an $S = 5/2$ system in an axial symmetry, such as molecules with Fe(III) ions in certain environments. The axial crystal field now removes the sixfold degeneracy down to three Kramers doublets as shown in Fig. 3.24. The spin Hamiltonian yields an energy separation between the $|\pm 1/2\rangle$ and $|\pm 3/2\rangle$ states equal to 2D and between the $|\pm 3/2\rangle$ and $|\pm 5/2\rangle$ states equal to 4D. The value of D depends on the nature of the ligands and how strongly they interact with the paramagnetic ion. We can distinguish two experimental limiting cases: when D is small $(D \ll h\nu)^4$ the Zeeman term is dominant leading to both

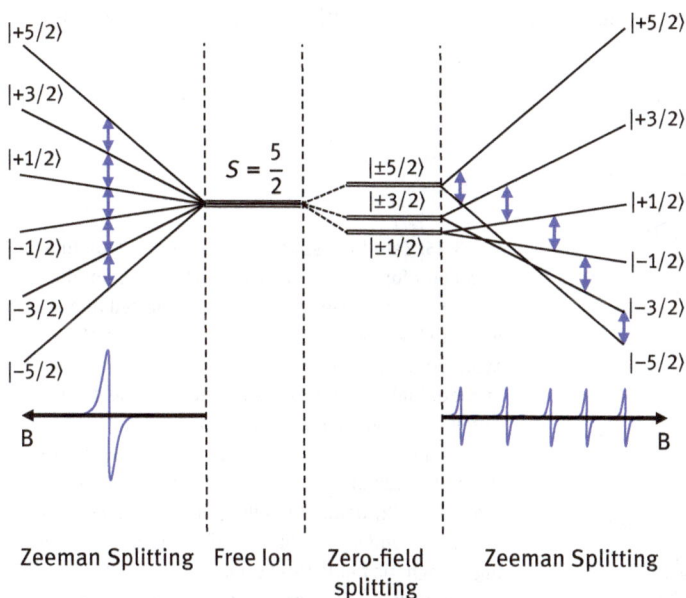

Zeeman Splitting Free Ion   Zero-field   Zeeman Splitting
splitting

Fig. 3.24: Zero-field splitting for an $S = 5/2$ ion in axial symmetry. The diagram to the left of the free ion shows the Zeeman splitting in the absence of zero-field splitting which results in five transitions according to the selection rule $\Delta m_s = \pm 1$ occurring at the same field value (blue arrows) that yield overlapping EPR lines. In the presence of a small zero-field splitting with $D > 0$, the energy manifold splits into three Kramers doublets (the order of the doublets is reversed when $D < 0$). For small values of D ($D \ll h\nu$), the Zeeman interaction is dominant and we could have five inter- and intradoublet transitions resulting in a spectrum with five EPR lines that do not overlap in their magnetic field position. The equal intensity lines displayed in this figure are possible at higher temperatures where the population of the states is approximately the same. For experiments carried out at low temperatures, the intensity of the EPR lines increases as we move from low to high field since the population of the states decreases as we go from $|-5/2\rangle$ to $|+5/2\rangle$.

---

4 For reference: a frequency equal to $\nu \approx 9.5$ GHz, which is typical in X-band EPR, corresponds to a quantum of energy equal to $h\nu \approx 0.3\,\mathrm{cm}^{-1}$.

inter- and intradoublet transitions as shown in Fig. 3.24, which result in a five-line EPR spectrum. For the case of large D ($D \gg h\nu$), only one EPR line is observed due to transitions within the ground $|\pm 1/2\rangle$ doublet, since only intradoublet transitions are allowed by the transition rules. This signal can be highly anisotropic as in the case of Fe(III) in biological molecules with heme structures.

When $E/D \neq 0$, the doublets shift in energy relative to their positions in axial symmetry. Small admixtures between the doublets are also introduced making intradoublet transitions possible within all doublets now as shown in Fig. 3.25. The resulting EPR spectrum consists of extra lines with g-values in addition to the axial values of 6 and 2 that occur when only transitions within the $|\pm 1/2\rangle$ doublet are allowed. The assignment of these experimental values is possible with the aid of rhombograms.

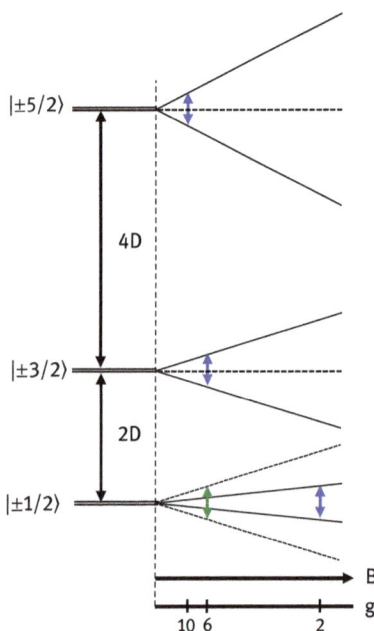

Fig. 3.25: Zero-field splitting for an $S = 5/2$ ion in axial symmetry for the case of large D ($D \gg h\nu$). Solid lines indicate energy levels with $\vec{B} \parallel z$ and dotted lines indicate levels with $\vec{B} \perp z$, where z is the axis of symmetry. Because of the large value of D only intradoublet transitions are possible. In the case of purely axial symmetry, the selection rules do not allow transitions within the $|\pm 3/2\rangle$ and $|\pm 5/2\rangle$ pure doublets. Transitions are only possible within the ground $|\pm 1/2\rangle$ doublet resulting in a single EPR line with $g_\parallel = 2$ and $g_\perp = 6$. (It should be noted that at very high magnetic fields ($B > 500$ mT), the $|\pm 1/2\rangle$ levels are no longer linearly split and there is a slight splitting of the $|\pm 3/2\rangle$ level. This behavior does not have any significant effect on most EPR spectra since they occur at fields less than 500 mT).

3.9.2 Rhombograms

For half-integer spin systems with $S > 1/2$ where the zero-field splitting removes the orbital degeneracy, the effective (observed) g-values can be calculated for each spin state and plotted versus the rhombicity ratio E/D to produce a series of graphs called a rhombogram. A rhombogram is used to determine the E/D value for a particular system by correlating the observed g-values on an EPR spectrum to the g-values on the rhombogram with the appropriate value of S. The range of the horizontal E/D coordinate axis runs between the values of 0 (axial symmetry) and 1/3 (maximum rhombic symmetry) and the vertical g-value axis runs through all possible values de-pending on the spin system (e.g., 0–6 for $S = 3/2$, 0 to 10 for $S = 5/2$, 0–14 for $S = 7/2$ and 0–18 for $S = 9/2$).

Rhombograms for $S = 3/2$ and $5/2$ are shown in Fig. 3.26. The graph for each state contains three curves that represent all the possible g-values for that state. The prin-cipal g-values for each state are obtained by drawing a line perpendicular to the E/D axis passing through a particular value of E/D. For example, for an $S = 3/2$ system in axial symmetry at low temperatures we will observe EPR lines with $g = 2$ and 4 as shown on the bottom plot in Fig. 3.26(a) for $E/D = 0$. When the temperature increases and the excited state gets populated, then a line with $g = 6$ will appear in the spectrum as shown in the top plot of Figure 3.26(a) for $E/D = 0$. If an experimental spectrum con-sists of lines with $g = 1.46$, 2.00 and 5.46, then it can be assigned to an $S = 3/2$ system in maximum rhombicity $E/D = 0.33$. Since these g-values are common to both doublets, assigning the lines to the correct doublet can be done by comparing the intensities of

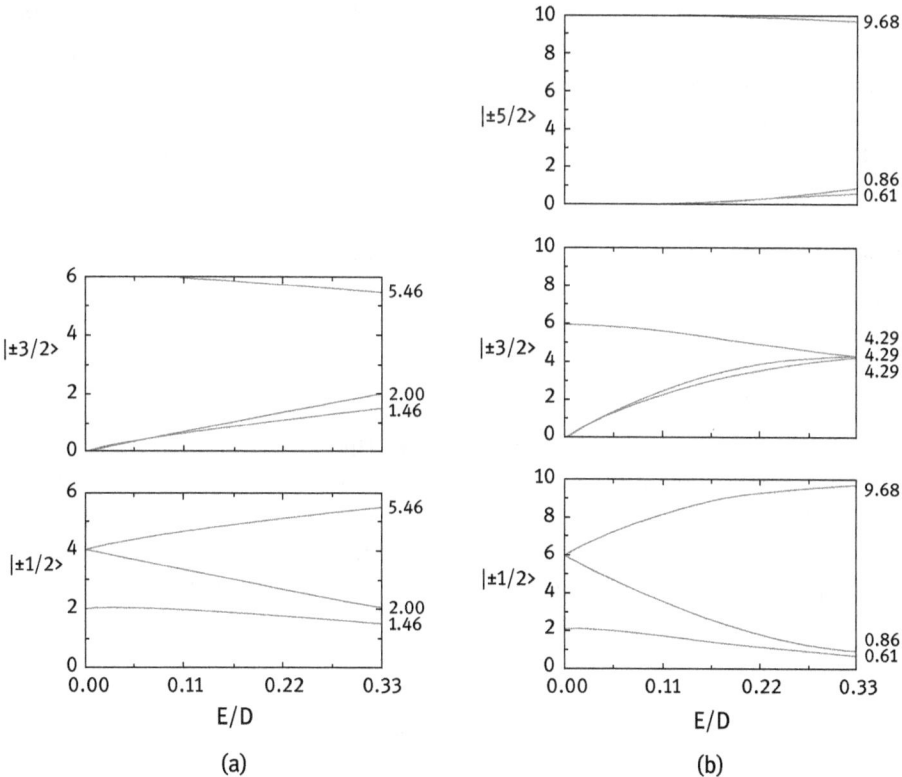

Fig. 3.26: Rhombograms for (a) $S = 3/2$ and (b) $S = 5/2$ systems. The vertical axis gives the g-value for the particular state while the horizontal axis gives the degree of rhombicity E/D. Similar rhomobograms can be generated for $S = 7/2$ and $S = 9/2$.

the lines as the temperature of the sample is changed: the lines corresponding to the $|\pm 1/2\rangle$ doublet will increase in intensity as the temperature is lowered. In the same way, a spectrum consisting of lines with $g = 2.0$ and 6.0 can be assigned to the $|\pm 1/2\rangle$ doublet of an $S = 5/2$ system according to the bottom plot in Fig. 3.26(b). A line at $g \approx 4.3$ can be assigned to the $|\pm 3/2\rangle$ doublet of an $S = 5/2$ system in extreme rhombic symmetry. This line is typically quite intense since all the resonances occur at the same g-value as indicated by the converging lines on the right vertical axis of the $|\pm 3/2\rangle$ rhombogram in Fig. 3.26(b). In the same way, any other lines present in the spectrum can be assigned to $S = 5/2$ states.

Figure 3.27 illustrates how rhombograms are used in the case of a number of biological $S = 5/2$ systems.

3.9.3 Integer spin (non-Kramers) systems

These systems produce spectra that are fundamentally different from the ones produced by half-integer spin systems since the level degeneracy can be completely removed by the zero-field splitting down to the level of singlets. In addition, in the presence of an external magnetic field the energy levels show a nonlinear variation with the magnetic field. There are cases where this splitting is larger than the microwave quantum at X-band frequencies (as stated earlier, at X-band: $v \approx 9.5$ GHz, $hv \approx 0.3$ cm^{-1}) making detection of an EPR signal impossible. These systems were inappropriately called "EPR silent" but they do in fact produce EPR signals when higher frequencies such as Q-band are used ($v \approx 34$ GHz, $hv \approx 1.2$ cm^{-1}). The $S = 1, 2$ systems are the most common in this group and will be discussed below.

3.9.3.1 $S = 1$
An example of an $S = 1$ system is the Ni(II) ion. This is a d^8 ion with two unpaired electrons in the d-orbital states according to Hund's rules, resulting in a total spin of 1. In an axially symmetric environment, the crystal field splits the degenerate triplet state (state multiplicity = $2S+1 = 3$) into a singlet $|0\rangle$ and a doublet $|\pm 1\rangle$ separated by an energy D. When a magnetic field $B_\parallel$ is applied parallel to the z-axis, the energies of the two states are given by

$$E_{|0\rangle} = 0 \tag{3.111}$$

$$E_{|\pm 1\rangle} = D \pm g_\parallel \beta B_\parallel \tag{3.112}$$

indicating that the $|\pm 1\rangle$ state is linearly split with respect to the magnetic field. When the magnetic field is in the xy-plane ($B_\perp$) and with $D \ggg g_\perp \beta B_\perp$ the energy levels are given by

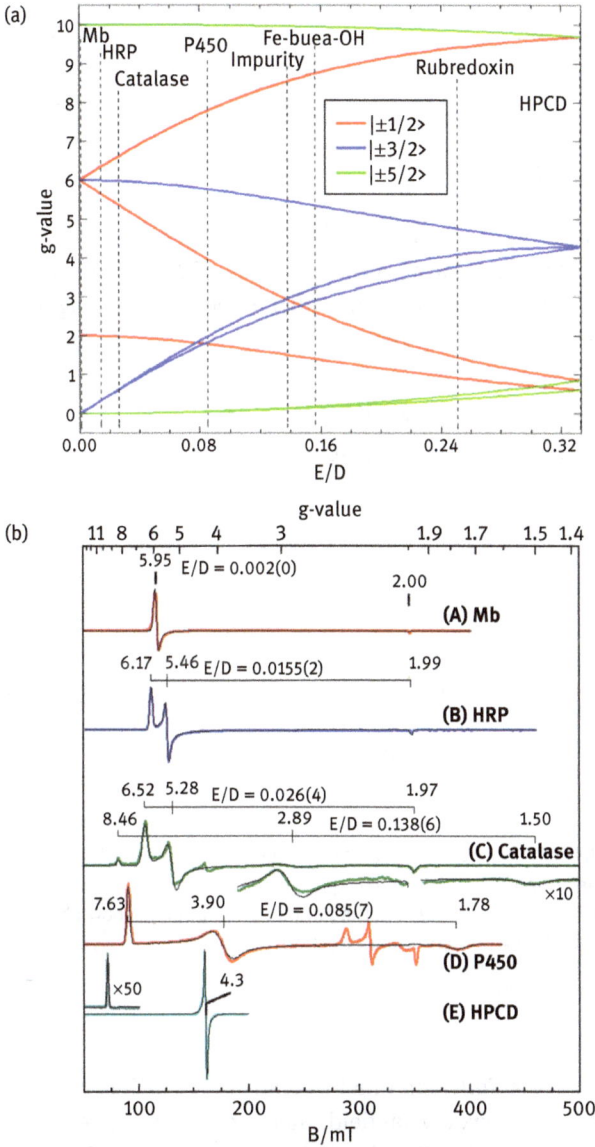

Fig. 3.27: Use of rhombograms in analyzing EPR signals. (a) $S = 5/2$ rhombogram that combines the effective g-values versus E/D plots for all three Kramers doublets into one plot. Each curve color corresponds to the state labeled with the same color in the inset. The vertical dashed lines indicate the E/D values that correspond to the observed g-values for the systems in (b). (b) EPR spectra (colored lines) and simulations (black lines on spectra) of (A) metmyoglobin, (B) horseradish peroxidase, (C) catalase, (D) cytochrome P450 and (E) homoprotocatechuate dioxygenase. Experimental conditions: $T \approx 10$ K, except (E) where $T \approx 2$ K, $v \approx 9.66$ GHz. The simulation parameters are for $S = 5/2$ with E/D values as listed. The values in parenthesis are the width $\sigma_{E/D}$ of the distribution in E/D [27].

$$E_{|0\rangle} \approx -\frac{g_\perp^2 \beta^2 B_\perp^2}{D} \tag{3.113}$$

$$E_{|-1\rangle} = D \tag{3.114}$$

$$E_{|1\rangle} \approx D + \frac{g_\perp^2 \beta^2 B_\perp^2}{D} \tag{3.115}$$

indicating that two of the states now vary quadratically with the magnetic field. This situation is illustrated in Fig. 3.28 for an arbitrary value of D. The magnetic anisotropy due to the zero-field splitting in this ion is also reflected in its magnetic susceptibility which has two values $\chi_\parallel$ and $\chi_\perp$ [26, 25].

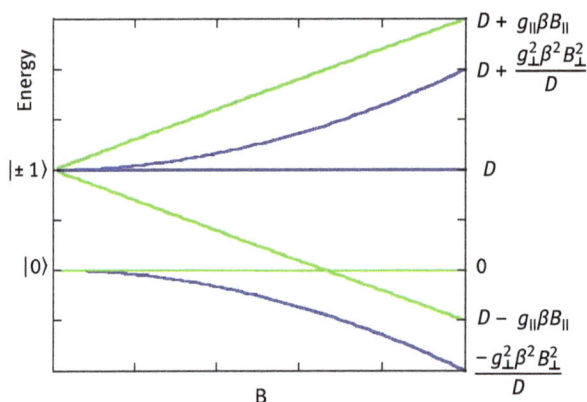

Fig. 3.28: Energy-level diagram of Ni(II) in the presence of an axial crystal field and an external magnetic field. The magnetic field is oriented along the z-axis (∥) or in the xy-plane (⊥). The behavior of the energy states when the magnetic field is in the parallel orientation is shown by the green lines and by the blue lines when in the perpendicular direction.

3.9.3.2 $S = 2$

An example of an $S = 2$ system is the Fe(II) ion in a low spin state. Fe(II) is a d^6 ion and in an octahedral symmetry in a weak crystal field, there are four unpaired electrons on the orbital states according to Hund's rules giving rise to a total spin equal to 2. The spin multiplicity is $2S+1 = 5$ giving rise to a quintet state. A zero-field Hamiltonian of the form

$$H = D\left(S_z^2 - 2\right) + E\left(S_x^2 - S_y^2\right) + \beta\vec{S} \cdot g \cdot \vec{B} \tag{3.116}$$

splits the quintet into a singlet $|0\rangle$ and two doublets $|\pm 1\rangle$ and $|\pm 2\rangle$. The separation between $|0\rangle$ and $|\pm 1\rangle$ is equal to D and between $|\pm 1\rangle$ and $|\pm 2\rangle$ is 3D as shown in Fig. 3.29. The zero field also removes the degeneracy of the $|\pm 1\rangle$ and $|\pm 2\rangle$ doublets producing small splittings $\Delta_1 = 6E$ and $\Delta_2 = 3E^2/D$ with E and D as defined in the above Hamiltonian. It is obvious that in a purely axial symmetry ($E = 0$), the two doublets remain degenerate but in the presence of distortions that introduce rhombic components into the crystal field the degeneracy is removed. The presence of a magnetic field B has no effect on the singlet state but the levels of the doublets $|\pm 1\rangle$ and $|\pm 2\rangle$ split quadratically as a function of B as shown. The resonance condition for such states with small zero-field splittings Δ is consequently more complicated than in the case of Kramers ions where such splittings are not possible and is given by

$$(h\nu)^2 = (g\beta B)^2 + \Delta^2 \tag{3.117}$$

It is obvious from this condition that when the microwave quantum $h\nu$ is smaller than Δ, there will be no EPR signal.

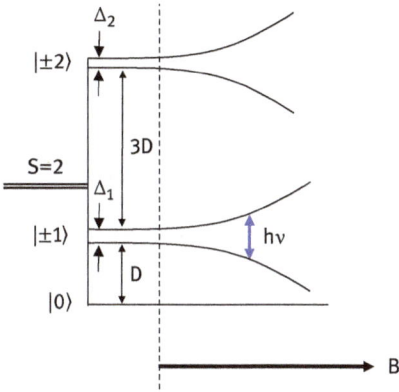

Fig. 3.29: Zero-field splitting of the degenerate quintet spin state of an $S = 2$ system such as Fe(II) into a singlet and two doublets. In the presence of rhombic distortions, the degeneracy of the doublets is completely removed giving rise to energy levels that vary quadratically in the presence of a magnetic field B.

3.10 Hyperfine interactions

Many transition series ions have nuclei with a nonzero spin angular momentum $\vec{I}$ that gives rise to a nuclear magnetic moment, which interacts with the magnetic moments of unpaired electrons within the ion. The electrons of a particular ion can also interact with the nuclei of other ions and with ligands that include atoms with nonzero nuclear spin. These interactions are called hyperfine interactions and give rise to hyperfine structure in the EPR spectra. The phenomenological explanation of how the electronic energy states split under the influence of the nuclear spin was briefly discussed in Section 2.9.

Quantum mechanically, $\vec{I}$ has similar behavior as the orbital angular momentum $\vec{L}$ with magnitude given by $|\vec{I}| = \sqrt{I(I+1)}\hbar$ and has discrete projections along the axis of quantization (z-axis) given by $I_z = m_I\hbar$ where the quantum number m_I varies from $-I$ to $+I$ in integer steps. Because of the space quantization of $\vec{I}$ the electron–nuclear interaction produces a splitting in the electronic energy levels and gives rise to hyperfine structure in the EPR spectrum. This was explored in Chapter 2 for the case of an $S = 1/2$ and $I = 3/2$ ion and can also be applied to the case of an $S = 1/2$ and $I = 3/2$ ion in an isotropic environment. Since m_I now has four discrete values, the nuclear spin $\vec{I}$ has four different orientations relative to the axis of quantization resulting in four hyperfine lines in the EPR spectrum as illustrated in Fig. 3.30. Because the strength of the dipolar interaction depends on the relative orientation between the two magnetic dipole moments, this contributes an anisotropic term to the hyperfine interaction.

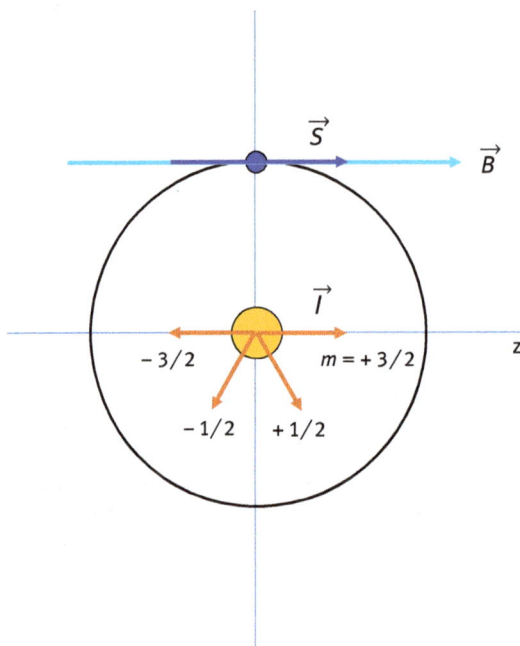

Fig. 3.30: Dipolar interaction between the magnetic moments of the electron and its nucleus that leads to hyperfine interaction in ions such as Cu(II). The axis of quantization z is taken in the same direction as the external magnetic field B. The nuclear spin I can have four different orientations relative to the z-axis resulting in four different projections along the axis of quantization leading to slightly different local fields experienced by the electron.

The hyperfine contribution to the spin Hamiltonian has the general form

$$\mathcal{H}_{\text{nuc}} = \vec{S} \circ \mathbf{A} \circ \vec{I} \tag{3.118}$$

where $\mathbf{A}$ is the hyperfine tensor. In isotropic systems, the hyperfine tensor is replaced by the hyperfine constant A. This interaction of a spin I nucleus with a spin S electron results in the splitting of the m_S electronic levels. The multiplicity of these levels is given by $(2I + 1)$ and each level is labeled by the quantum number m_I

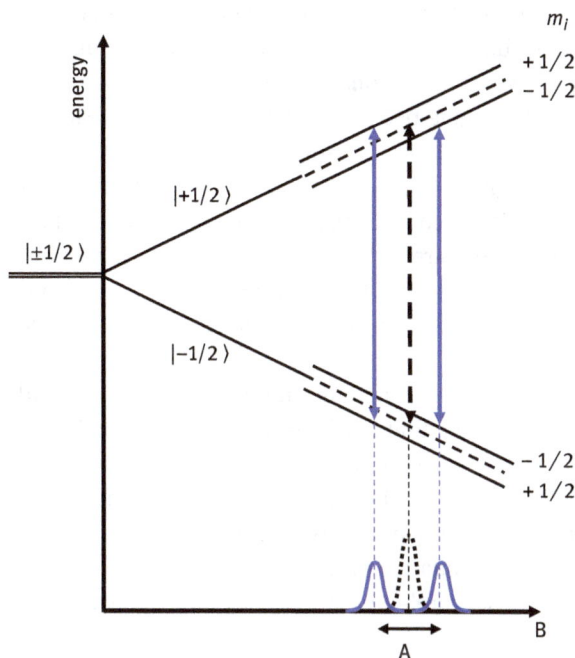

Fig. 3.31: Hyperfine splitting for an ion with $S = 1/2$ and $I = 1/2$. In the absence of the hyperfine interaction only one EPR line is observable (dotted trace), but when the hyperfine splitting occurs two hyperfine lines result with a separation equal to the hyperfine constant A. The value of the hyperfine constant A is dependent on the spin density of the electron at the nucleus and therefore on bonding in the molecule. An electron in an orbital that is mainly metal based will have high spin density at the nucleus of the metal and a large hyperfine constant (roughly >50 MHz). If the metal electron interacts with nearby ligands, the spin density and hyperfine constant are much less.

(m_I takes values from $-I$ to $+I$ in integer steps) as shown in Fig. 3.31 for an $S = 1/2$ system with a nuclear spin of $I = 1/2$.

The complete Hamiltonian that includes both the Zeeman and the interaction between the electronic spin and the nuclear magnetic moment can now be written as

$$\mathcal{H} = \beta \vec{S} \cdot g \cdot \vec{B} + \vec{S} \circ A \circ \vec{I} \tag{3.119}$$

In the limiting high-field case where A is much smaller than $g\beta B$, it is shown in Chapter 6 that to first-order perturbation the following energies can be calculated:

$$E = \pm \frac{1}{2} g\beta B \pm \frac{A}{4} \tag{3.120}$$

We see now that the two original states that existed in the absence of hyperfine interaction are now each split into two states each according to the m_I values as illustrated in Fig. 3.31 for $S=1/2$. Transitions are now allowed between the m_I levels.

Since the spin angular momentum of an emitted photon is equal to $\pm \hbar$, conservation of angular momentum dictates that these transitions involve changes in angular momentum equal to 1, only allowing transitions that obey the selection rules: $\Delta m_S = \pm 1$ and $\Delta m_I = 0$. These transitions give rise to $(2I+1)$ equally spaced EPR lines of equal intensity. For an $S = 1/2$ and $I = 1/2$ ion, the original single EPR line is now split into two hyperfine lines. The magnetic field separation of the hyperfine lines is given by the hyperfine constant A which is characteristic of the particular ion. Typical transition series ions with nuclear spin are Cu^{2+} ($I = 3/2$), Mn^{2+} ($I = 5/2$) and Co^{2+} ($I = 7/2$). Figure 3.32 illustrates the case of an $S = 1/2$ and $I = 3/2$ ion in an isotropic environment which gives rise to four hyperfine lines of the same intensity and separation A.

The hyperfine lines all have the same intensity, and the separation between adjacent lines is equal to A. It should be noted that not all isotopes of a particular element have a nonzero nuclear spin. For example, of the four stable and naturally abundant isotopes of Fe (^{54}Fe, ^{56}Fe, ^{57}Fe, ^{58}Fe) only ^{57}Fe, with an abundance of 2.12%, has nuclear spin ($I = \frac{1}{2}$). The intensity of the hyperfine lines in a particular sample also depends on the natural abundance of the isotope, which in cases of very low abundances might not even be visible in the EPR spectrum, as is the case with Fe-containing samples.

3.10.1 Superhyperfine interactions

Hyperfine interactions are also possible between a paramagnetic ion and nearby ligands possessing nuclear spin such as ^{14}N with $I = 1$. These so-called superhyperfine interactions produce groups of EPR lines of varying intensities. The interaction of the central ion with n equivalent ligand nuclei produces $(2nI + 1)$ EPR lines. An example of this kind of interaction is given by Ingram [6] for the case of a free radical electron interacting with protons on CH, CH_2 and CH_3 groups. In the CH case, the delocalized electron interacts with one proton resulting in a two-line hyperfine structure as shown in Fig. 3.31. In the CH_2 case, the electron interacts with two protons. If the strength of the interaction is of the same magnitude for both protons, then the hyperfine levels from the interaction with the first proton will further split into two levels each in a symmetric manner as shown in Fig. 3.33. This will result in essentially three energy levels since the –1/2 level of the second nucleus produced from the splitting of the +1/2 level of the first nucleus coincides in energy with the +1/2 level of the second nucleus produced from the splitting of the –1/2 level of the first nucleus in both the +1/2 and –1/2 spin levels. This will give rise to three transitions resulting in three hyperfine lines with the centerline having double the intensity since the transition is between doublet states with double the population of the other states where hyperfine transitions occur. Consequently, the intensity ratio of the hyperfine lines follows a binomial expansion as 1:2:1 which follows from the

(a)

(b)

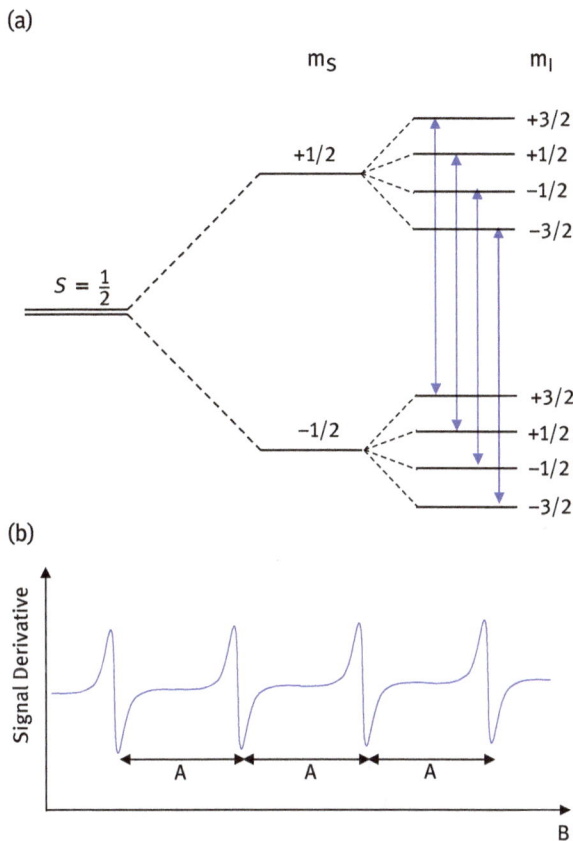

Fig. 3.32: (a) Hyperfine splittings in a system with $S = \frac{1}{2}$ and $I = 3/2$. Allowed transitions between the nuclear levels obey the selection rule $\Delta m_S = \pm 1$ and $\Delta m_I = 0$ and are indicated by the arrows. (b) Ideal EPR hyperfine spectrum for a species with $I = 3/2$. The g-value is determined from the field position of the center of the spectrum.

coefficients of the expansion of $(a + b)^2$. When the electron interacts with three protons as in the case of the CH_3 group, each of the final nuclear levels in Fig. 3.33 will be again split into two levels. If the strength of the interaction is the same for all protons, then the levels will symmetrically split in the manner shown in Fig. 3.34. Due to the overlap of intermediate levels as before, we only have four distinct energy levels and, consequently, four hyperfine lines. Due to the overlap of three levels in the intermediate states, the intensity of the middle lines will be three times as strong as the intensities of the outside lines as shown in Fig. 3.34. The intensity ratio of the lines again follows a binomial expansion as 1:3:3:1.

This process can be repeated for the interaction of one electron equally interacting with any number of protons. For an interaction with n protons, the result will be $(n + 1)$ hyperfine lines with an intensity ration given from the coefficients of the

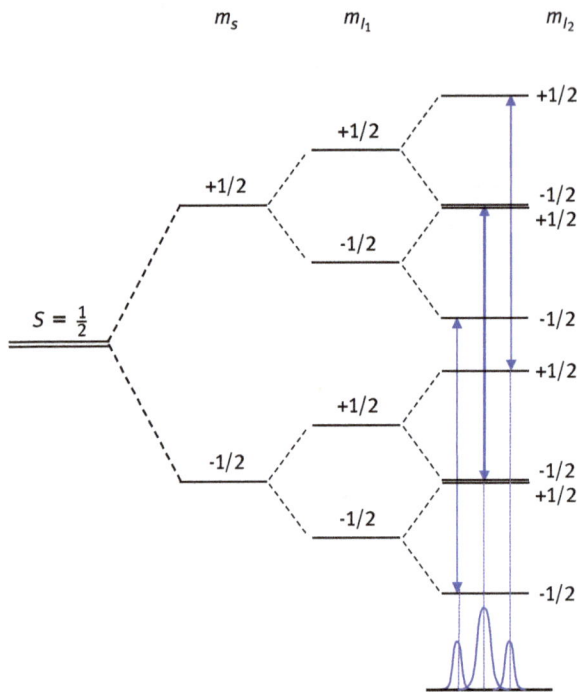

Fig. 3.33: Interaction of an electron ($S = 1/2$) with two protons ($I_1 = I_2 = 1/2$) such as the ones found in CH_2 groups. Central hyperfine line is more intense since the transition that gives rise to this line is between two states where two nuclear levels come together to form each state. The population of these levels is consequently larger than the other nuclear levels resulting in a more intense hyperfine line.

binomial expansion of $(a + b)^n$. The method just described of a free radical electron interacting with protons of organic groups can also be applied to the case where the electron is bound to a paramagnetic ion in a molecule and interacts with the protons of the ligands of the paramagnetic ion. In cases where $I \neq 1/2$, the intensity of the hyperfine lines will no longer follow a binomial distribution but a more complicated intensity ratio results.

In the case where the strength of the interactions between the electron and the nuclei do not have the same magnitude, the symmetric splittings described in the examples above will not happen since now the hyperfine levels will be split by different amounts. The hyperfine lines described above will now split into more components. For example, the overlapping energy levels in Fig. 3.33 will no longer be degenerate resulting in the central line splitting into two lines giving a four-line hyperfine spectrum. In general, the stronger interaction is considered first producing $(2I_1 + 1)$ hyperfine lines. Each of these lines is further split into $(2I_2 + 1)$ components due to the weaker interaction with the second nucleus. The field separation of these

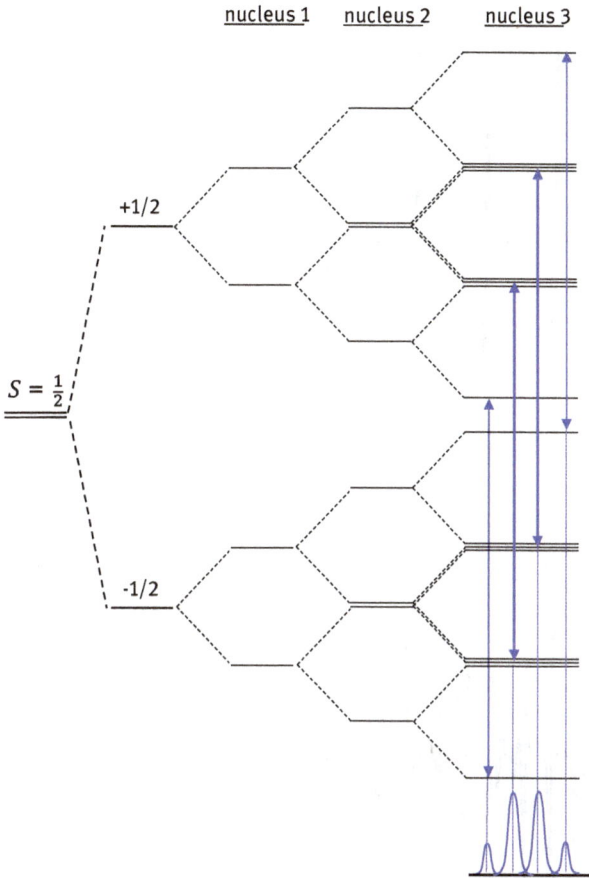

Fig. 3.34: Interaction of an electron ($S = 1/2$) with three protons ($I_1 = I_2 = I_3 = 1/2$) such as the ones found in CH_3 groups. Interactions of the electron with the three protons are of equal strength resulting in symmetrically split energy levels that give rise to hyperfine lines with intensities having the ration 1:3:3:1.

lines is smaller than the separation of the lines due to the first nucleus. Figure 3.35 illustrates this for the case of the benzene [28] and pyrazine [29] radical anions.

3.10.2 Resolution of hyperfine spectra

The above example of the EPR spectrum of the pyrazine radical anion illustrates an issue with complicated hyperfine spectra where many overlapping hyperfine lines are present in one spectrum. In such a case, especially when the lines are numerous and with substantial broadening, it might not be possible to correctly make assignments for the lines. When this happens, it is advisable to run the sample at a lower

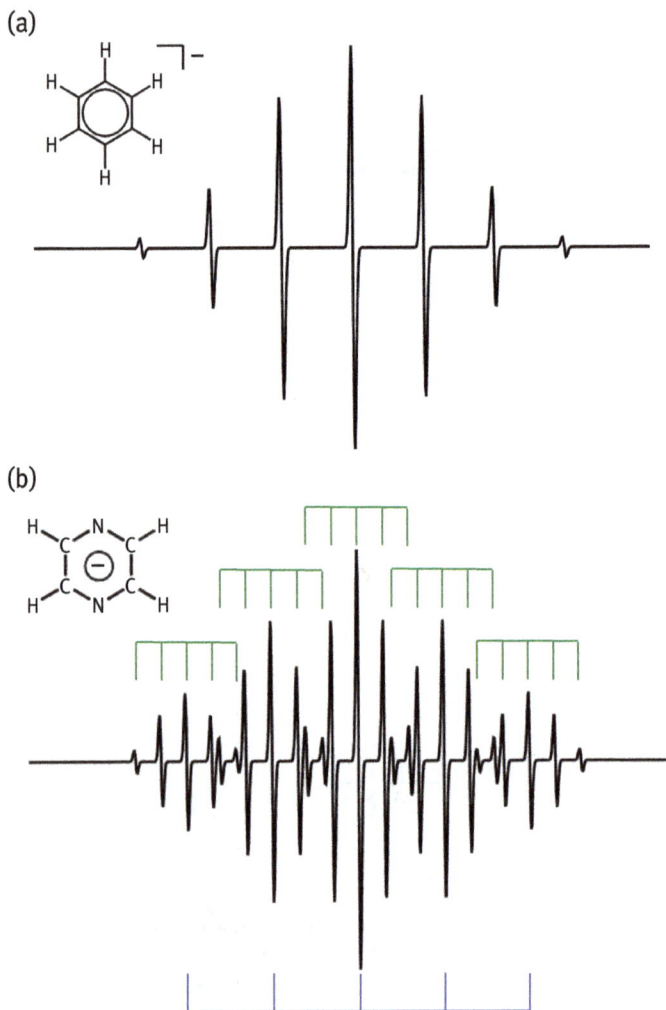

Fig. 3.35: (a) EPR spectrum of the benzene radical anion. The delocalized electron is free to move within the benzene ring and interacts equally with all six protons ($I = 1/2$) producing seven hyperfine lines ($n + 1$) with the intensity ratio 1:6:15:20:15:6:1. (b) EPR spectrum of the pyrazine radical anion. The delocalized electron interacts strongly with the two equivalent nitrogens ($I = 1$) producing a five-line hyperfine spectrum with relative intensities 1:2:3:2:1 indicated by the blue ladder. A weaker interaction with the more distant four equivalent protons ($I = 1/2$) splits each of the five lines into a further quintet with the intensity ratio 1:4:6:4:1 indicated by the green ladders.

frequency to increase the field spacing between the hyperfine lines as illustrated in Fig. 3.36. Typically EPR spectra are collected at X-band frequencies, and a shift to S-band frequencies can produce a good resolution of overlapping hyperfine lines. Bruker's Elexsys series of spectrometers allow for quick switch between X- and S-band frequencies as will be discussed in Chapter 6.

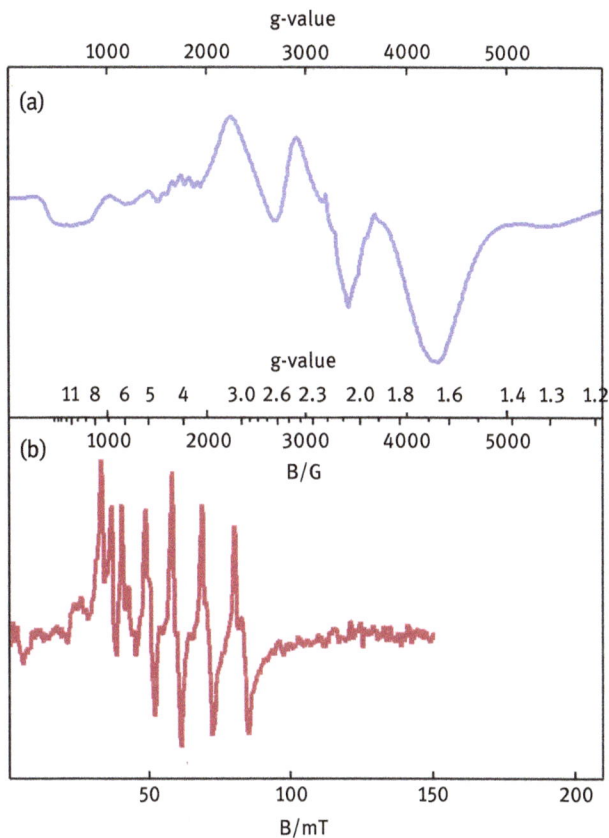

Fig. 3.36: Experimental spectra of Mn(II)EDTA (EDTA, ethylenediaminetetraacetic acid) at (a) X-band ($v = 9.65$ GHz) and (b) S-band ($v = 3.48$ GHz). The six-line hyperfine spectrum of Mn(II) with $I = 5/2$ is not completely resolved at X-band frequencies but it is completely resolved at S-band frequencies. Both spectra were taken at a microwave power of 0.3 mW and at a temperature $T = 20$ K.

3.10.3 Hyperfine anisotropy

The contributions to hyperfine structure can be divided into two types: the dipolar interaction between the magnetic moment of the electron and the nucleus that we have discussed above, and the Fermi contact interaction due to the nonzero value of the electronic orbital wave function at the nucleus. The first contribution gives rise to anisotropic values of A and the second contribution gives rise to isotropic A. We briefly discussed the anisotropic contribution in Section 2.9 and both contributions will be further discussed in Chapter 6.

When hyperfine anisotropy is present, the separation of the hyperfine lines can change as the magnetic field is rotated relative to the symmetry axes of a crystal.

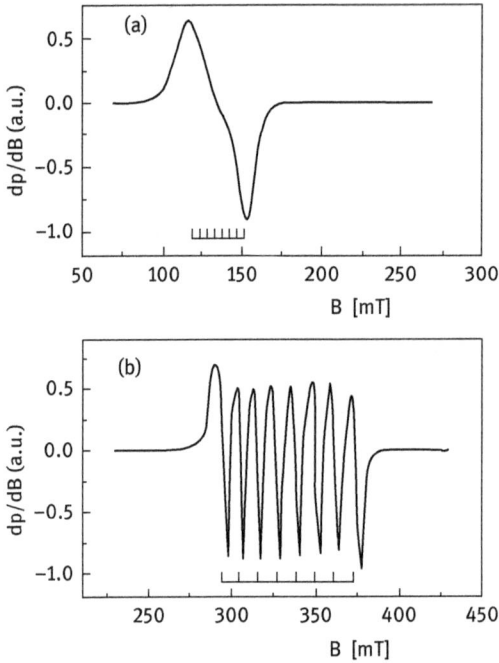

Fig. 3.37: X-band spectra of Co(II) doped in SrLaAlO$_4$ at $T = 20$ K. (a) $\vec{B} \perp$ [001] direction and (b) $\vec{B} \parallel$ [001] direction. $g_\parallel = 1.975$, $g_\perp = 4.870$, $A_\parallel = 104 \times 10^{-4}$ cm^{-1}, $A_\perp = 101 \times 10^{-4}$ cm^{-1} [30].

When only g-value anisotropy is present, the hyperfine spectrum will move as a whole during rotations, but when both are present, the spectrum will move and the separation between the hyperfine lines will also change. In certain single crystals where both g-value and hyperfine anisotropy are present, the hyperfine lines can shift in such a way upon rotation of the magnetic field relative to crystalline axes of symmetry that they start to overlap and produce a broad asymmetric line. Such a case is illustrated in Fig. 3.37 for a crystal of SrLaAlO$_4$ doped with Co(II). The eight hyperfine lines due to the ^{59}Co isotope ($I = 7/2$) are partially resolved when $\vec{B}$ is parallel to the [001] direction but are unresolved when $\vec{B}$ is in the perpendicular direction resulting in a broad asymmetric line. In the latter case, this is due to the value of $A_\perp$ being smaller than $A_\parallel$. The field position of the spectrum as a whole changes in the two different orientations of $\vec{B}$ due to the anisotropy in the g-value.

In the case of polycrystalline spectra, in the presence of both g-value and hyperfine anisotropy, each of the features of the spectra shown in Fig. 3.21 will be split into hyperfine lines. Figure 3.38 shows what happens to Fig. 3.21 when the electron interacts with a nucleus with $I = 1/2$.

In the case of Cu(II), the electron interacts with an $I = 3/2$ nucleus to produce a four-line hyperfine spectrum. In an axial symmetry, the g-value is anisotropic with $g_\parallel = 2.32$ and $g_\perp = 2.07$ with an anisotropic hyperfine constant with negligible $A_\perp$

(a)

(b)

(c)

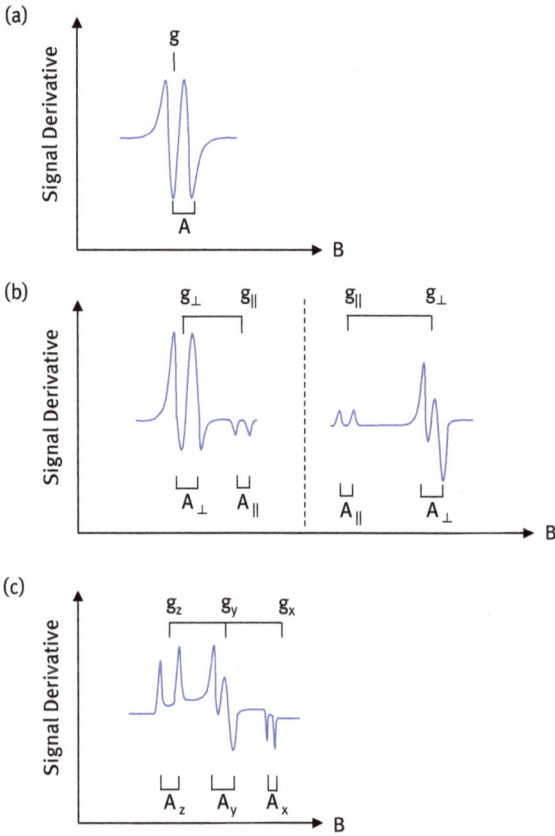

Fig. 3.38: Powder EPR spectra of a spin $S = 1/2$ interacting with a nucleus with $I = 1/2$ in various symmetries: (a) isotropic, (b) axial and (c) rhombic.

and $A_\parallel = 146$ G as shown in Fig. 3.39. (NB: in recent literature, hyperfine constants are quite often quoted in cm^{-1} using the conversion formula $A(cm^{-1}) = g\beta A(G)$ where $\beta = 4.66864 \times 10^{-5}\ cm^{-1}\ G^{-1}$ or in MHz using $A(MHz) = 1.3996\ gA(G))$ [31].

3.11 Symmetry effects

This section focuses on the effects that the surrounding ligands have on the paramagnetic ion within the context of the anisotropy of the g-value. The effect on the g-value can be quite pronounced since these interactions can greatly influence the energy levels of the electron. The spatial arrangement of the surrounding molecules or ligands can give rise to different point symmetries experienced by the paramagnetic ion. Crystal field theory approaches these interactions from the point of view of electrostatic interactions between the paramagnetic ion and the interacting

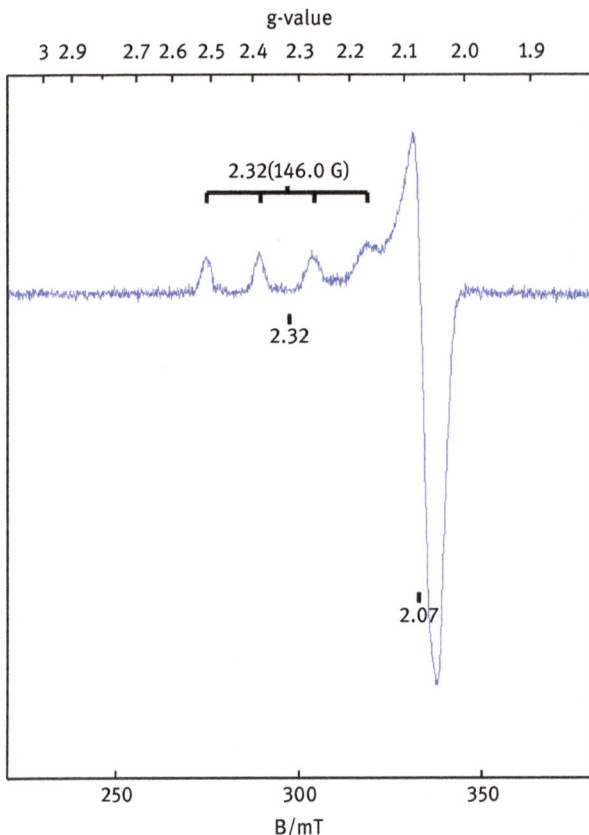

Fig. 3.39: X-band spectrum of a frozen solution of 1 mM CuEDTA. Frequency = 9.657 GHz, microwave power = 2θ W, temperature = 30 K. The four-line hyperfine structure with $A_\parallel$ = 146 G is centered at $g_\parallel$ = 2.32. $A_\perp$ is too small to be resolved from within the broad asymmetric line at $g_\perp$ = 2.07.

entities. These can be smaller molecules with large electric dipole moments such as the water molecules in the rare earth ethylsulfates and triflates or ligands coordinating transition series ions. The geometry of these ligands has big influences on the EPR spectra as demonstrated in the previous section and is shown again in Fig. 3.40, where the single-crystal spectra of an S = 1/2 ion are displayed along with the three major geometries that are encountered in EPR. Crystal field theory is important for paramagnetic ions bound in molecules and crystal lattices and will be discussed in more detail in Chapter 4.

The simplest geometry, other than spherical, is simple cubic where the ligands are equidistant from the paramagnetic ion, and the interactions between ligands and paramagnetic ion are of equal strength. Since all directions in space are equivalent in this case, the result of an EPR experiment is a single line since it does not matter how the magnetic field is oriented relative to the molecular axes. The g-value is isotropic and has one unique value as shown in Fig. 3.40(a).

When the arrangement of the ligands is such that there is a unique direction in space, it can be represented by a tetragonal geometry with a unique direction along the z-axis. In such a case, the ligand environment is the same in the xy-plane but different along the z-axis. We call this symmetry "axial," and the EPR signal will now shift as the magnetic field is rotated from the z-axis to the xy-plane producing signals at two different field values. Typically, the unique axis is taken as the "parallel direction" (∥) and the other direction as the "perpendicular direction" (⊥). The two values of the g-value are labeled as $g_\parallel$ and $g_\perp$ as shown in Fig. 3.40(b). The $g_\perp$ line is more intense since it consists of spin packets originating from resonances along both the x- and y-axes that overlap at the same magnetic field value due to x and y being equivalent axes.

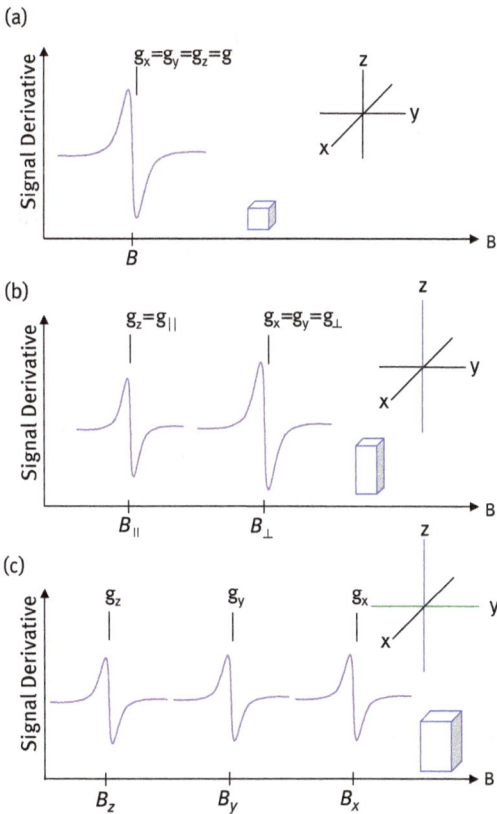

Fig. 3.40: Single-crystal EPR spectra for an $S = 1/2$ ion in (a) isotropic, (b) axial and (c) rhombic crystal symmetries. The crystal symmetries are indicated by the coordinate systems and the geometric shapes that accompany the spectra.

For ligand arrangements where there are no equivalent directions in space, the geometry can be represented by an orthorhombic structure (all lower symmetry geometries, such as hexagonal and trigonal, have the same effect as orthorhombic). In such a case, there will be EPR signals at different magnetic field values for orientations of the magnetic field along the x-, y- and z-axes. We now have a "rhombic" symmetry with three different g-values that characterize the paramagnetic ion: g_x, g_y and g_z as illustrated in Fig. 3.40(c). The relative intensities of the lines depend on the number of molecules having a specific orientation relative to the coordinate axes.

In the case of polycrystalline or "powder" samples[5], the EPR signals are not as straightforward as the signals from crystal samples displayed in Fig. 3.41. With crystal samples, the magnetic field can be set in a particular direction and the crystal can be rotated to place its principal axes at any possible orientations relative to the field. Alternatively, in spectrometers where the magnet can be rotated, the crystal is kept at a constant orientation, and the magnetic field is rotated relative to the principal axes of the crystal. In the case of powder samples, the microcrystals that make up the sample are randomly oriented in space so that the EPR spectrum now consists of a summation of all the possible orientations of the microcrystals relative to the magnetic field. In the isotropic case, the symmetry is the highest and all orientations in space are equivalent, causing the EPR signals from all microcrystals to overlap producing one single EPR line as shown in Fig. 3.40(b), top. In the axial symmetry case, there will be two extreme g-values at $g_\parallel$ and $g_\perp$ with all other possible resonances occurring between these two extreme values. Adding the contributions from all these signals (think of them as spin packets that add up in a similar way to the one displayed in Fig. 3.12) produces the powder spectrum illustrated in Fig. 3.41(b), middle. For the case of the rhombic symmetry, there will be signals at every value of the magnetic field between the extrema g_z and g_x that result in the powder spectrum shown in Fig. 3.41(b), bottom. The shapes of this first derivative spectra for the powder axial and rhombic spectra are easier to understand if we first consider the lines that result from absorption spectra and then take the first derivative. This will be discussed in more detail in Chapter 4.

This chapter attempted to present the basic shapes of EPR signals that one expects to encounter when running EPR experiments. It is obvious from the discussion in this section that a more thorough presentation of the effects of symmetry on the EPR spectra is needed. This will be done in the next chapter, where a more in-depth discussion of the effects of symmetry on EPR spectra will be presented.

---

[5] Biological samples are typically dissolved in a liquid buffer and then frozen, resulting in frozen solution samples which have a behavior similar to powder samples.

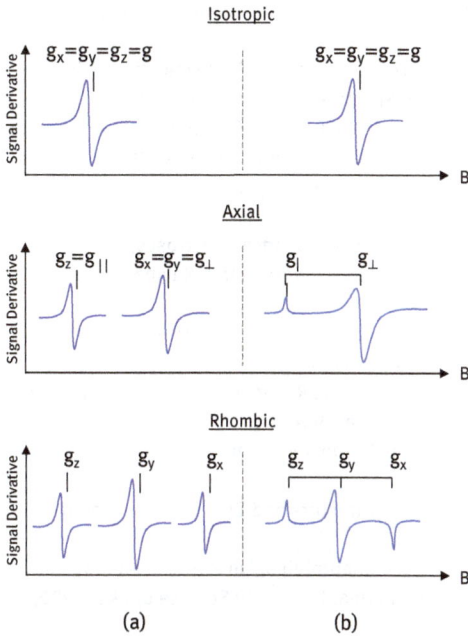

Isotropic

$g_x=g_y=g_z=g$ $g_x=g_y=g_z=g$

Axial

$g_z=g_{||}$ $g_x=g_y=g_\perp$ $g_{||}$ $g_\perp$

Rhombic

g_z g_y g_x g_z g_y g_x

(a) (b)

Fig. 3.41: EPR signals for an $S = 1/2$ system from (a) a single-crystal and (b) polycrystalline sample in environments with isotropic, axial and rhombic symmetries. The spectra for the polycrystalline sample contain microcrystallites with random orientations in space which result in signals for every field value between the extreme values of the single-crystal spectra.

References

[1]   C. Corvaja, "Introduction to Electron Paramagnetic Resonance," In Electron Paramagnetic Resonance, M. R. Brustolon, Ed., Hoboken, John Wiley & Sons, 2009, pp. 3–35.
[2]   L. Sorin and M. V. Vlasova, Electron Spin Resonance of Paramagnetic Crystals, New York, London, Plenum Press, 1973.
[3]   P. Rieger, Electron Spin Resonance: Analysis and Interpretation, Cambridge, GBR: Royal Society of Chemistry, 2007.
[4]   B. Bleaney and A. Abragam, Electron Paramagnetic Resonance of Transition Ions, Oxford University Press, 1970.
[5]   C. Benelli and D. Gatteschi, Introduction to Molecular Magnetism: From Transition Metals to Lanthanides, Weinheim, Wiley-VCH, 2015.
[6]   D. J. E. Ingram, "Electron Spin Resonance," *Handuch der Physik*, vol. XVIII, p. 94–144, 1968.
[7]   O. S. Leifson and C. D. Jeffries, "Dynamic Polarization of Nuclei by Electron-Nuclear Dipolar Coupling in Crystals," *Physical Review*, 122, p. 1781–1795, 1961.
[8]   C. P. Poole, Electron Spin Resonance: A Comprehensive Treatise on Experimental Techniques (2nd ed.), Dover, 1996.
[9]   M. Bersohn and J. Baird, An Introduction to Electron Paramagnetic Resonance, New York, W. A. Benjamin, Inc., 1966.

[10] J. Weil, J. Bolton and J. Wertz, Electron Paramagnetic Resonance: Elementary Theory and Practical Applications, New York, Wiley-Interscience, 1994.

[11] W. Low, Paramagnetic Resonance in Solids, New York and London, Academic Press, 1960.

[12] G. W. Brudvig, "Electron Paramagnetic Resonance Spectroscopy," In Methods in Enzymology, vol. 246, K. Sauer, Ed., San Diego, Academic Press, 1995, pp. 536–554.

[13] B. J. Hales, "Intrinsic and Extrinsic Paramagnets as Probes of Metal Clusters," In Methods in Enzymology, vol. 227, J. F. Riordan and B. L. Vallee, Eds., San Diego, Academic Press, 1993, pp. 384–395.

[14] J. H. Golbeck and A. van der Est, "Electron Paramagnetic Resonance Spectroscopy," In Molecular Biophysics for the Life Sciences, N. Allewell, L. O. Narhi and I. Rayment, Eds., New York, Springer, 2013, pp. 175–213.

[15] J. W. Orton, Electron Paramagnetic Resonance: An Introduction to Transition Group Ions in Crystals, New York, Gordon and Breach Science Publishers Inc., 1969.

[16] J. A. Weil, J. R. Bolton and J. E. Wertz, Electron Paramagnetic Resonance: Elementary Theory and Practical Applications, New York, Wiley-Interscience, 1994.

[17] B. R. McGarvey, "ESR of Transition Metal Complexes," *Transition Metal Chemistry*, 3, 1967, pp. 89–199.

[18] E. I. Solomon and A. B. P. Lever, Inorganic Electronic Structure and Spectroscopy (vol. I), New York, John Wiley & Sons, 1999.

[19] G. Palmer, "Electron Paramagnetic Resonance of Metalloproteins," In Physical Methods in Bioinorganic Chemistry, L. Que, Ed., Sausalito, California, University Science Books, 2000, pp. 121–185.

[20] B. Bleaney, "Magnetic Resonance Spectroscopy and Hyperfine Interactions," In Handbook on the Physics and Chemistry of Rare Earths, vol. 11, K. A. Gschneider, Jr. and L. Eyring, Eds., Amsterdam, Elsevier Science Publishers, 1988, pp. 323–407.

[21] D. T. Petasis, G. Bellesis, N. Vander Ven and S. Friedberg, "Electron Paramagnetic Resonance of Ce3+ in La(CF3SO3)3.9H2O," *Journal of Applied Physics*, 70, 1991, pp. 5998–6000.

[22] G. Palmer, "The Electron Paramagnetic Resonance of Metalloproteins," *Biochemical Society Transactions*, 13, p. 548–560, 1985.

[23] W. R. Hagen, Biomolecular EPR Spectroscopy, Boca Raton, CRC Press, 2008.

[24] G. Palmer, "Electron Paramagnetic Resonance of Hemoproteins," In Iron Porphyrins, A. B. P. Lever and H. B. Gray, Eds., New York, Addison-Wesley, 1983, pp. 43–88.

[25] O. Kahn, Molecular Magnetism, New York, VCH Publishers, 1993.

[26] R. L. Carlin, Magnetochemistry, Berlin, Springer-Verlag, 1986.

[27] D. T. Petasis and M. P. Hendrich, "Quantitative Interpretation of Multifrequency Multimode EPR Spectra of Metal Containing Proteins, Enzymes, and Biomimetic Complexes," In Methods in Enzymology, vol. 563, Amsterdam, Elsevier, 2015, pp. 171–208.

[28] J. R. Bolton, "13C Hyperfine Splitting in the Benzene Negative Ion," *Molecular Physics*, 6(2), p. 219–221, 1963.

[29] A. Carrington and J. Dos Santos-veiga, "Electron Spin Resonance Spectra of Nitrogen Heterocyclic Radical Ions," *Molecular Physics*, 5(1), p. 21–29, 1962.

[30] P. Aleshkevych, M. Berkowski, W. Ryba-Romanowski and H. Szymczak, "EPR and Optical Spectra of Cobalt in SrLaAlO4," *Physica Status Solidi (B)*, 218, p. 521–526, 2000.

[31] A. Bencini and D. Gatteschi, "Electron Paramagnetic Resonance Spectroscopy," In Inorganic Electronic Structure and Spectroscopy, vol. I, E. I. Solomon and A. B. P. Lever, Eds., New York, John Wiley & Sons, 1999, pp. 93–159.

[32] J. G. Castle and D. W. Feldman, "Electron Spin-Lattice Relaxation of Chromium in MgO," *Physical Review*, 121, p. 1349, 1961.

4 Symmetry and crystal field theory

The EPR spectra of solid metal compounds are greatly influenced by the coordination environment of the paramagnetic ion which causes shifts in the energy states of the ion. The "ligands" that make up the coordination environment of transition metal and rare earth ions are typically negatively charged ions/small molecules or neutral molecules arranged in specific geometries about the paramagnetic ion. EPR is very sensitive to the local symmetry experienced by the paramagnetic ion, allowing the detection of small distortions in the symmetry of its neighbors even though the overall molecule or crystal appears not to be distorted. This property makes EPR a powerful tool in determining structural details of the ligands in the metal neighborhood. In order to understand how the ligand coordination influences the EPR spectra, we must consider how ligands influence the energy states of the paramagnetic ion, which is the focus of this chapter.

The electrons of interest in transition series ions occupy d-orbitals and in rare earth ions occupy f-orbitals. The angular probability densities of these orbitals are shown in Fig. 2.3 along with the s- and p-orbitals and reproduced in Fig. 4.1. The interaction between the electrons on d- and f-orbitals of metal centers in complexes

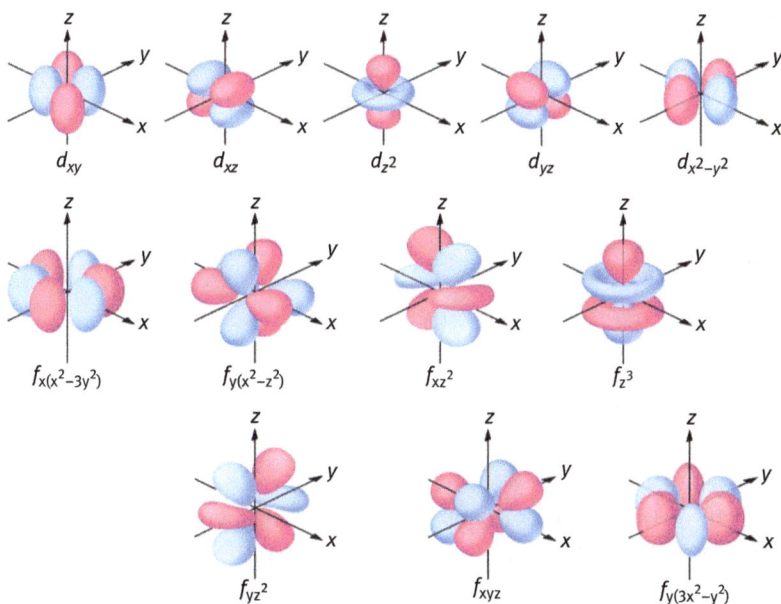

Fig. 4.1: Orbital diagrams of valence states for transition series elements. Three-dimensional polar plots of the angular dependence of the wave functions for d- and f-orbitals. (Image reproduced from "Chemistry: Atoms First 2e" published by OpenStax and can be accessed for free at https://openstax.org/books/chemistry-atoms-first-2e/pages/1-introduction).

https://doi.org/10.1515/9783110417562-004

and the electrons of the surrounding ligands give rise to differences in the energies of the orbitals of the metal ion as compared to that of the free metal ion.

4.1 Effect of symmetry on the orbital states of transition series ions

It is obvious from the angular distribution of d- and f-orbitals shown in Fig. 4.1 that depending on which orbital a particular electron occupies, it will interact differently with the ligands that are arranged in a certain geometry around the metal ion. The simplest ligand geometry that we can consider is the spherical symmetry where the negatively charged ligands are symmetrically distributed on the surface of a sphere that is centered on the metal ion. Since the ligands are interacting equally with the metal electrons, the net effect on the electron energies is to raise the energy of all electrons by an equal amount relative to the energy of the electrons in the free ion. The highest symmetry we typically observe in metal complexes is the octahedral symmetry (O_h) where the metal ion is placed at the center of coordinates with the ligands placed symmetrically about the origin along each coordinate axis as shown in Fig. 4.2. The ligand coordinates are $(x, y, z) = (\pm a, 0, 0)$, $(0, \pm a, 0)$ and $(0, 0, \pm a)$.

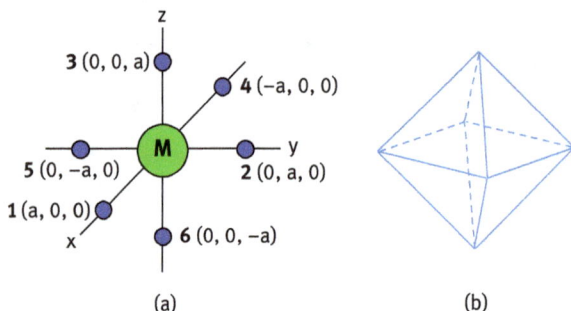

Fig. 4.2: (a) Arrangement of the six ligands (blue) about a metal ion (green) in octahedral symmetry. (b) A regular octahedron. The ligands are equidistant from the metal ion, placed at the corners of the octahedron with the metal at the center of gravity.

To see the effect of the ligand geometry on the metal orbitals, consider the d_{xy} and $d_{x^2-y^2}$ orbitals of a transition series metal ion. If we look at the projection of these orbitals onto the xy-plane, as shown in Fig. 4.3, we observe that electrons on these orbitals are going to interact differently with the octahedrally arranged ligands: the electron on the $d_{x^2-y^2}$ orbital will have a higher energy than the electron on the d_{xy} orbital since it gets closer to the ligand ion. Electrons on the d_{xz} and d_{yz} orbitals have the same geometric arrangement relative to the ligand ion; consequently, they will have their energies shifted by the same amount as the electrons on the d_{xy}

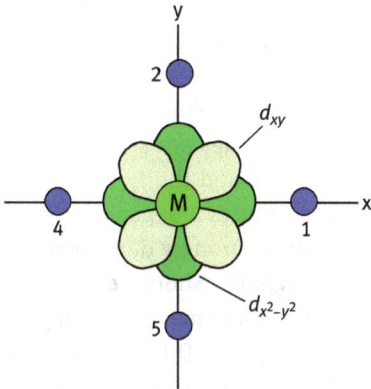

Fig. 4.3: Metal orbitals and octahedrally arranged ligands on the xy-plane. The projections of the d_{xy} and $d_{x^2-y^2}$ orbitals of the metal onto the xy-plane are shown.

orbital. In the same way, electrons on the d_{z^2} orbital will have their energies shifted by the same amount as the electrons on the $d_{x^2-y^2}$ orbital.

Comparing the interaction energies with those of the spherically symmetric ligands where the radius of the sphere is equal to a (the distance between ligands and metal ion in the O_h symmetry), we see that the energies of the $d_{x^2-y^2}$ and d_{z^2} orbitals are going to be shifted to a higher level and those of the d_{xy}, d_{xz} and d_{yz} orbitals will be shifted to a lower level. This situation is illustrated in Fig. 4.4. The triplet orbital state that is formed by the d_{xy}, d_{xz} and d_{yz} orbitals is given the designation t_{2g}, which comes from the name given to this symmetry in group theory. The

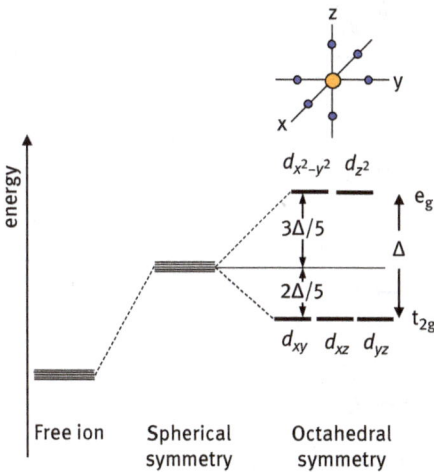

Fig. 4.4: The effects of spherical and octahedral crystal field symmetries on the energies of the d-orbital levels. Spherical symmetry causes a shift of the orbital energy levels without removing the degeneracy of the states. In octahedral symmetry, the orbital degeneracy is removed, resulting in a lower energy triplet t_{2g} and a higher energy doublet e_g. These two states are separated by an energy Δ that depends on the strength of the crystal field.

group theoretical designation of the $d_{x^2-y^2}$ and d_{z^2} orbitals is e_g, as indicated in Fig. 4.4. The shift of the e_g and t_{2g} states relative to the spherically symmetric states is not symmetric since the electrons on the $d_{x^2-y^2}$ and d_{z^2} orbitals are significantly closer to the ligands than the electrons on the d_{xy}, d_{xz} and d_{yz} orbitals. If the energy separation between the e_g and t_{2g} states is Δ, then the e_g states are shifted by an amount $3\Delta/5$ higher than the spherically symmetric states and the t_{2g} states are shifted by an amount $2\Delta/5$ in the opposite direction. The value of Δ depends on how strong the crystal field is. The strength of the crystal field is strongly dependent on the ligand to metal distance and on the nature of the ligands. Shorter distance a means stronger electrostatic interactions between the ligand and metal electrons resulting in a strong crystal field that in turn results in a larger Δ. Different types of ligands have different electronegativities which result in interactions of different strength between the ligand and metal electrons resulting in different values of Δ.

The orbital energy states can further be affected by changes in the geometry of the ligands around the metal. Deviations from octahedral symmetry will result in further removal of the degeneracies that exist in octahedral symmetry. For example, a tetragonal distortion where the two ligands along the z-axis are shifted further away from the metal ion will result in the complete removal of the degeneracy of the e_g states and raise the d_{xy} state a little higher than the d_{xz} and d_{yz} states. A further reduction in symmetry by also displacing the ligands along the y-axis results in the complete removal of all remaining degeneracies. These two examples are illustrated in Fig. 4.5.

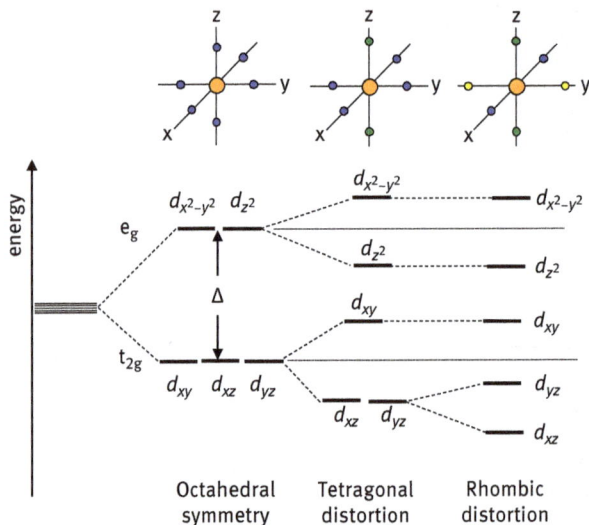

Fig. 4.5: The effects of axial and rhombic distortions on the splitting of the orbital states that result from an octahedral crystal field symmetry. Lower symmetries result in the removal of more degeneracies of the orbital levels. In tetrahedral symmetry, splitting of orbitals is the same as in O_h but t_{2g} and e_g are flipped around with smaller Δ.

4.2 Spin states of transition series ions

4.2.1 Spin-only systems

The removal of the degeneracy of the orbital states under the action of a crystal field gives rise to the different spin states that a particular ion can have. For transition series ions, the splitting of the d states into a t_{2g} triplet and an e_g doublet separated by an energy Δ can yield different values of the net spin depending on how large Δ is compared to the thermal energy $k_B T$. If the crystal field is not very strong and $\Delta \sim k_B T$, then the electrons have enough thermal energy to access the excited e_g states. If, on the other hand, $\Delta \ll k_B T$ the electrons do not have enough energy to access the e_g states but have to distribute themselves on the t_{2g} states instead and might have to pair up yielding a different total spin state.

This can be illustrated by considering systems with iron in one of its more common ionic forms: Fe(II) and Fe(III). Iron is very important in solid state and biological systems which makes it an ideal example to explore the interactions of the metal with its environment. The Fe(III) ion has five d-electrons (d^5). In the case of a weak octahedral crystal field, the five electrons will be able to access all orbital states. According to Hund's rules they will occupy one orbital state each to reduce their energy. We now have one orbital with one electron of spin up each. The total spin of this ion is then equal to $S = 5/2$ implying that the ion possesses a large magnetic moment. This is called the "high-spin" state (hs state). If, on the other hand, the crystal field is strong so that the e_g levels are not thermally accessible, all five electrons must distribute themselves on the t_{2g} states, subject to the Pauli exclusion principle. Since there are five electrons and only three available states, two of the states will be occupied by paired electrons and one state by one unpaired electron. Since paired electrons have zero net spin, then the net spin of the ion is now equal to $S = 1/2$. This is called the "low-spin" state (ls state) of Fe(III). This situation is illustrated in Fig. 4.6 along with Fe(II) which has six d-electrons (d^6) yielding hs and ls states of two and zero, respectively. In the same way, we can obtain the spin states for other ions such as Mn(II) and Ru(III) both of which are d^5 ions with spin states $S_{hs} = 5/2$ and $S_{ls} = 1/2$.

4.2.2 Systems with spin–orbit coupling

For these systems, we have to distinguish between strong and weak spin–orbit coupling. In some $3d$ ions, the orbital angular momentum is not completely quenched and some weak spin–orbit coupling occurs. In $4d$ and $5d$ ions, the spin–orbit coupling is more substantial due to the larger atomic numbers. For rare earth ions, the $4f$ electrons are shielded by the filled $5s$ and $5d$ orbitals decreasing the influence of the crystal field and allowing strong spin–orbit coupling. In the case of weak

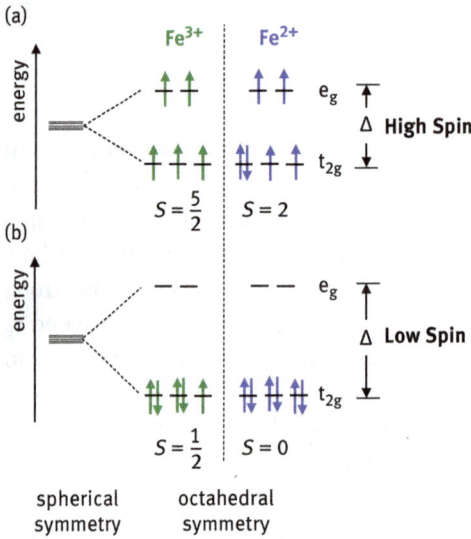

Fig. 4.6: Spin states of Fe(III) and Fe(II) ions in a crystal field of octahedral symmetry. (a) When Δ is small, all orbital states are thermally accessible to the electrons allowing the electrons to occupy all states subject to the exclusion principle. The result is a high-spin state for both ions. (b) For large Δ, the e_g states are not thermally accessible to the electrons and they have zero occupancy. The electrons all occupy the t_{2g} states, again subject to the exclusion principle, resulting in a low-spin state.

spin–orbit coupling, the crystal field acts on the orbital states first causing the splitting discussed in the previous section and spin–orbit acts as a perturbation on the split orbital states. The effect of spin–orbit coupling on the energy-level splitting of a ^2D system in the presence and absence of an octahedral crystal field is shown in Fig. 4.7.

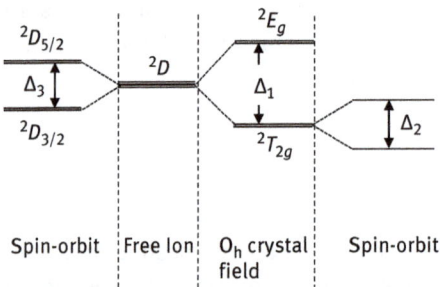

Fig. 4.7: Energy-level diagram for an ion with L and S contributions to the angular momentum leading to a ^2D energy manifold ($L = 2$ and $S = 1/2$ yielding $J = 3/2, 5/2$). The effect of spin–orbit coupling on the energy states in the absence and presence of an octahedral crystal field are shown. $\Delta_1 = 10Dq$, $\Delta_2 = 3\zeta/2$, $\Delta_3 = 5\zeta/2$ [1].

We have seen in Chapters 2 and 3 that the spin–orbit Hamiltonian is given by

$$\mathcal{H}_{so} = \lambda \vec{L} \cdot \vec{S} \tag{4.1}$$

where for ground states $\lambda = \pm \zeta/2S$ with the plus sign for less than half-filled and minus for more than half states. The spin–orbit coupling coefficient ζ depends on the element and some values are shown in Tab. 2.1.

Spin–orbit coupling has an effect on the g-value of states with quenched orbital angular momentum since it can couple excited states and the ground state leading to a partially unquenched orbital angular momentum in this state which shifts the g-value away from 2.0.

4.2.3 Kramers and non-Kramers systems

The introduction of spin states also allows us to distinguish between ions with an odd number of unpaired electrons, resulting in a net half-integer spin $S = 1/2, 3/2, 5/2, \ldots$ called *Kramers ions* and ions with an even number of electrons resulting in integer spin $S = 1, 2, 3, \ldots$ called *non-Kramers ions*. These two broad categories of spin systems have fundamentally different behaviors inside a crystal field and in the presence of an external magnetic field. Figure 4.8 illustrates these differences for the case of $S = 5/2$ and $S = 2$ ions. The $S = 5/2$ ion splits into three individually degenerate doublets in the absence of an external magnetic field. For a general spin S, the doublets are $|\pm m_S\rangle$ states. When a magnetic field B is turned on, each doublet will split linearly versus B, and spin transitions can be induced between the states according to the EPR resonance condition and provided the selection rule $\Delta m_S = \pm 1$ is satisfied (Fig. 4.8(a)). As shown in Section 3.7, the intensity of transitions within a doublet is proportional to

$$I \propto \left| \left\langle +m_S \left| \vec{B}_1 \cdot \vec{S} \right| -m_S \right\rangle \right|^2 = \left| \left\langle +m_S \left| B_{1x}S_x + B_{1y}S_y \right| -m_S \right\rangle \right|^2 \tag{4.2}$$

The quantization axis is always parallel to $\vec{B}$, so the only component of $\vec{B}_1$ that contributes to the signal intensity is $\vec{B}_1 \perp \vec{B}$.

The $S = 2$ ion is split into a singlet and two doublets. For general spin S, the doublets are $|\pm m'\rangle$ states which are symmetric and antisymmetric combinations of the $|\pm m_S\rangle$ states:

$$\left| \pm m' \right\rangle = \frac{1}{\sqrt{2}} (|+m_S\rangle \pm |-m_S\rangle) \tag{4.3}$$

The two doublets are not always degenerate in the absence of a magnetic field but can sometimes (in lower than axial symmetries) have a small splitting Δ which is much smaller than the separation between the doublets. When a magnetic field is turned on, the doublets do not split in a linear fashion like the $S = 5/2$ doublets but instead have a quadratic dependence on B (Fig. 4.8(b)). The resonance condition differs substantially from the simple resonance condition for Kramers ions $hv = g\beta B$ which in this case is given by

$$(hv)^2 = \Delta^2 + (g_{\mathrm{eff}}\beta B \cos\theta)^2 \tag{4.4}$$

where $g_{\mathrm{eff}} \approx 2m_S g$ and θ is the angle between the magnetic field B and the quantization axis z as defined by the molecular axis. The selection rule that applies here is $\Delta m_S = 0$. The intensity of transitions within a doublet is proportional to

$$I \propto \left|\left\langle +m'_S\left|\vec{B}_1 \cdot \vec{S}\right| -m'_S\right\rangle\right|^2 = \left|\left\langle +m'_S\left|B_{1z}S_z\right| -m'_S\right\rangle\right|^2 \tag{4.5}$$

which means that only components of $\vec{B}_1 \parallel \vec{B}$ contribute to the signal intensity. It should be noted here that the direction of the quantization axis z is determined by the molecular or crystal environment of the spin rather than $\vec{B}$, resulting in signal intensities that are strongly dependent on molecular orientation.

It is obvious from this discussion that as long as the resonance condition and the selection rule are satisfied, Kramers ions will always produce EPR signals.[1] This is not always true for non-Kramers ions, since there can be instances where the zero-field doublet splitting Δ can be larger than the microwave quantum ($\Delta > hv$),

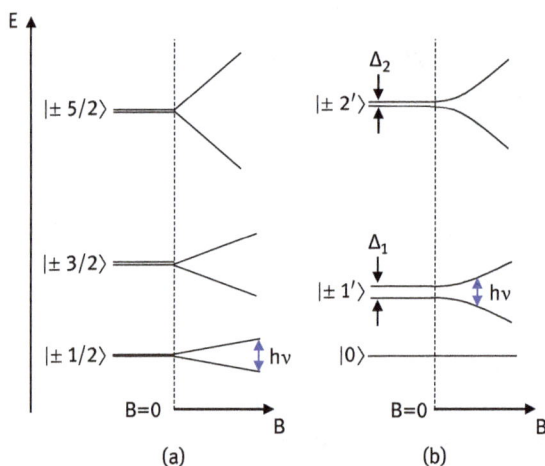

Fig. 4.8: Comparison of energy-level diagrams of (a) Kramers ($S = 5/2$) and (b) non-Kramers ($S = 2$) ions. In the presence of symmetries lower than axial, the doublet states $|\pm 1'\rangle$ and $|\pm 2'\rangle$ of the non-Kramers ion exhibit small zero-field splittings Δ_1 and Δ_2, respectively. The blue double arrows indicate possible EPR transitions.

---

1 There are certain Kramers systems with $g_z \gg g_{x,y}$, which have very small transition probabilities where an EPR signal is not observed.

thereby preventing transitions from ever happening. Consequently, the observation of EPR signals from non-Kramers ions requires spectrometers that operate at frequencies higher than the typical X-band frequencies of most commercial spectrometers [2]. Because of the splitting Δ, EPR signals from non-Kramers ions typically have large g-values since the resonances occur at low magnetic fields, as shown in Fig. 4.8(b). As $\Delta \to h\nu$, the resonance field approaches zero.

The different behaviors of Kramers and non-Kramers ions are due to their different time-reversal properties. In Kramers ions, the crystalline electric field lifts the degeneracy of the manifold down to spin doublets giving rise to the so-called Kramers doublets labeled as $|\pm m_s\rangle$. This is a result of Kramers' theorem [3], which applies to odd electron systems with states that are symmetric under time reversal in the presence of pure electric fields: the energy states are at least doubly degenerate. In other words, for each state $|\psi\rangle$, there is a time-reversed state $|\psi'\rangle = T|\psi\rangle$ with the same energy, where T is the time-reversal operator.[2] Such spin states split in a linear manner in the presence of an applied magnetic field $\vec{B}$, the resonance condition $h\nu = g\beta B$ applies here, and spin transitions can be induced between the $|+m_s\rangle$ and $|-m_s\rangle$ states when $\vec{B}_1 \perp \vec{B}$, provided the selection rule $\Delta m_s = \pm 1$ is satisfied as we have already seen in Chapter 3. For obvious reasons, conventional EPR spectroscopy was developed to study Kramers ions since their spectroscopy is pretty straightforward.

In non-Kramers ions, the time-reversed eigenfunctions do not have to be degenerate in the presence of an electric field. It is possible for the crystal field to completely remove the degeneracy of all states. A pair of energy levels of these spin systems has zero-field splitting Δ and a quadratic energy splitting as a function of an applied field $\vec{B}$. Early theories proposed that strains that lower the symmetry at the paramagnetic ion are responsible for the zero-field splitting Δ [4]. If these random distortions are the cause of the Δ splitting, then Δ can be expressed as

$$\Delta = \sqrt{\Delta_x^2 + \Delta_y^2} \tag{4.6}$$

where Δ_x and Δ_y have a distribution of values that are assumed to be Gaussian in nature and centered at zero. In an effective spin 1/2 formalism ($S_{\text{eff}} = 1/2$), the spin Hamiltonian, when the magnetic field is along the z-axis, can then be written in the form

$$\mathcal{H} = \beta g_z B_z S_z + \left(\Delta_x S_x + \Delta_y S_y\right) \tag{4.7}$$

---

2 During time reversal, the time variables describing the time evolution of a system become negative, reversing the direction of flow of time. The time-reversal operator T maps $t \to -t$. When T operates on angular momentum, it reverses its sign.

with eigenvalues

$$E = \pm \frac{1}{2}\left[(\beta g_z B_z)^2 + \Delta^2\right]^{\frac{1}{2}}\tag{4.8}$$

which allowed the interpretation of early integer-spin spectra that displayed unusually broad and asymmetric lineshapes. Even though this Hamiltonian was successful in interpreting the EPR spectra, it is physically inaccurate due to the fact that it predicts a finite spin expectation value even when $B_z = 0$.

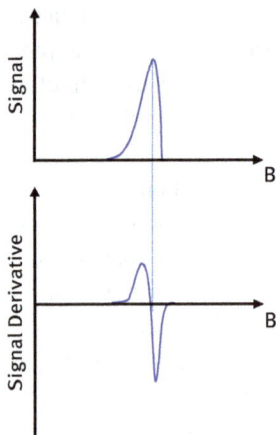

Fig. 4.9: A qualitative example of an EPR lineshape for transitions in a non-Kramers doublet, in the $S_{eff} = 1/2$ formalism, when there is an accidental degeneracy and the zero-field splitting is produced by random strains that have a Gaussian distribution [4]. The absorption line has a distorted Gaussian shape that produces a highly asymmetric first derivative shape. The same type of lineshape is obtained in the formalism where the actual spin S is used and the D and E terms have a distribution of values. In many non-Kramers systems, especially in metalloproteins, there is hardly any positive contribution to the signal derivative.

A more accurate way to interpret integer-spin spectra is to use the true spin S and the corresponding m_s states instead of treating each doublet as an effective spin 1/2 state. We can then use the Hamiltonian

$$\mathcal{H} = D\left[S_z^2 - \frac{S(S+1)}{3}\right] + E\left(S_x^2 - S_y^2\right) + \beta \vec{S}\cdot\hat{g}\cdot\vec{B}\tag{4.9}$$

where D and E are the axial and rhombic zero-field splitting parameters, respectively. The Zeeman term is assumed to be much smaller than D. In the absence of a rhombic term ($E = 0$), the states are degenerate. In the presence of a rhombic term, the degeneracy is removed, and the separation of the states increases as E increases. The asymmetric lineshapes result when the parameters D and E are allowed to have Gaussian distributions (skew-symmetric distributions actually yield better results) [5]. The effect of zero-field splitting is illustrated in Fig. 4.10 for the example of a spin $S = 2$ system, where in the presence of axial symmetry the spin manifold splits into a singlet, $|\pm 0'\rangle = |\pm 0\rangle$, and two doublets:

$$|\pm 1'\rangle = \frac{1}{\sqrt{2}}(|+1\rangle \pm |-1\rangle)\tag{4.10}$$

$$\left|\pm 2'\right\rangle = \frac{1}{\sqrt{2}}(\left|+2\right\rangle \pm \left|-2\right\rangle) \qquad (4.11)$$

In the absence of an external magnetic field, the two doublets split as a function of the ratio E/D as shown. The insets show the splitting of the two doublets versus an external magnetic field when $E/D \neq 0$.

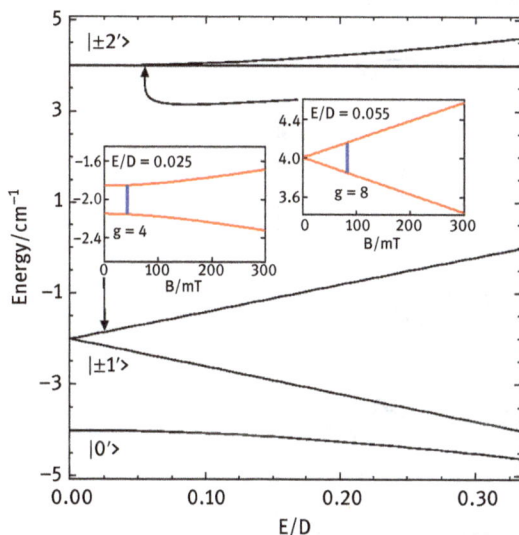

Fig. 4.10: Plot of energy versus E/D for an $S = 2$ system with $D = 2$ cm^{-1}. The doublets $\left|\pm 1'\right\rangle$ and $\left|\pm 2'\right\rangle$ are degenerate in the absence of a rhombic distortion ($E = 0$). The splitting of the two doublets as a function of the magnetic field along the molecular z-axis is shown in the two insets for two different values of the E/D ratio: 0.025 ($\left|\pm 1'\right\rangle$) and 0.055 ($\left|\pm 2'\right\rangle$). The blue lines indicate the EPR transitions when $v = 9.3$ GHz (reproduced with permission from [6]).

4.2.4 Examples of EPR spectra in non-Kramers systems

4.2.4.1 Single crystal of Fe(II) doped in zinc fluorosilicate

The ferrous ion in this complex is in a hs state with $S = 2$ [2]. The energy levels are split as shown in Fig. 4.8(b) and are parameterized using the spin Hamiltonian given by eq. (4.9). EPR signals obtained at Q-band frequencies (34 GHz) and low temperatures ($T = 10$ K) are shown in Fig. 4.11 for rotations of the external magnetic field $\vec{B}$ relative to the crystalline axis of symmetry. The EPR signal has a highly asymmetric lineshape as predicted by the qualitative shape shown in Fig. 4.9. The EPR signal originates from the first excited double $\left|\pm 1\right\rangle$ which is at an energy equal to D above the ground singlet $\left|0\right\rangle$ and has a small zero-field splitting equal to $\Delta_1 = 3E$ and eigenfunctions given by

$$\left|\pm 1'\right\rangle = \left|m_s = +1\right\rangle \pm \left|m_s = -1\right\rangle \qquad (4.12)$$

The second excited doublet $|\pm 2\rangle$ is at an energy equal to 3D above the first excited doublet and its degeneracy is also removed at zero field by an amount $\Delta_2 = 3E^2/D$. Simulations of the experimental signal at $\theta = 0°$ using the Hamiltonian of eq. (4.9) are shown by the dashed line in the inset of Fig. 4.11 with the fit parameters $D = 14.3$ cm^{-1}, $E = 0$ and $g_z = 2.415$. The simulations were created using the software SpinCount developed by Prof. Michael Hendrich of Carnegie Mellon University for the analysis of EPR spectra. A description of SpinCount can be found in reference [6].

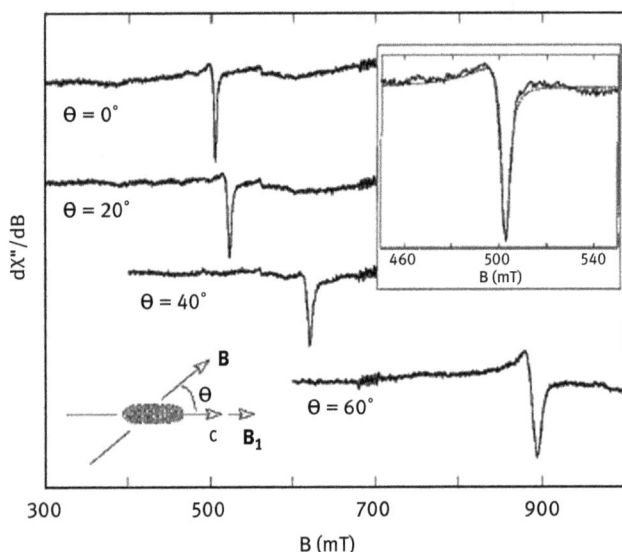

Fig. 4.11: EPR spectra of a 10 mg crystalline sample of Fe(II) in zinc fluorosilicate taken at Q-band frequencies (34 GHz) at a temperature of 10 K. The first signal ($\theta = 0°$) was taken with the external magnetic field $\vec{B}$ oriented parallel to the microwave magnetic field $\vec{B}_1$, which is also parallel to the c-axis of the crystal. Three more signals are shown for rotations of $\vec{B}$ relative to the c-axis as defined in the diagram within the plot. The inset shows enlarged detail of the signal at $\theta = 0°$ (solid line) with a superimposed simulation (dotted line). Image reproduced with permission from reference [2].

4.2.4.2 Single crystal of Pr(III) doped in La trifluoromethanesulfonate

Some of the first EPR experiments on non-Kramers ions were carried out in the 1950s by the group of Brebis Bleaney at the Clarendon Laboratory of Oxford University. The focus was on trivalent ions of the rare earth series where, starting with Ce(III), Kramers ions alternate with non-Kramers ions. Extensive work on the rare earth ethylsulfates included experiments on trivalent praseodymium doped in the diamagnetic host yttrium ethylsulfate (YES). The ethylsulfates exhibit a C$_{3h}$ point symmetry at the rare earth ion which is susceptible to local strains that lower the symmetry, especially when a paramagnetic ion substitutes for the diamagnetic ion of slightly different ionic radius, as Pr(III) does when it substitutes for Y(III) in YES. The free Pr(III) ion has a $^3H_4$ ground state and its only stable isotope has a nuclear spin $I = 5/2$. The nine-fold degeneracy of the $J = 4$ manifold is removed in the presence of a crystal field with C_{3h} symmetry resulting in three doublet and three singlet states [4]. The ground doublet is accidentally degenerate with the two states conjugated by time reversal and given by

$$|J, J_z\rangle = a|J = 4, J_z = \pm 2\rangle - b|J = 4, J_z = \mp 4\rangle \tag{4.13}$$

where the admixture coefficients are related according to $b \ll a$. The resulting spectrum consisted of six lines with asymmetric shape as expected if random distortions lift the accidental degeneracy of the ground state. The spectra are shown in Fig. 4.12. Similar spectra were observed when Pr(III) was doped in lanthanum ethylsulfate (LaES) and in the isostructural complex lanthanum trifluoromethanesulfonate (LaTFMS). The spectra were analyzed using the effective spin 1/2 Hamiltonian of eq. (4.7). The zero-field splitting Δ of the ground doublet was determined to be equal to 0.11 cm^{-1} in YES [7], 0.19 cm^{-1} in LaES [8] and 0.16 cm^{-1} in LaTFMS [9].

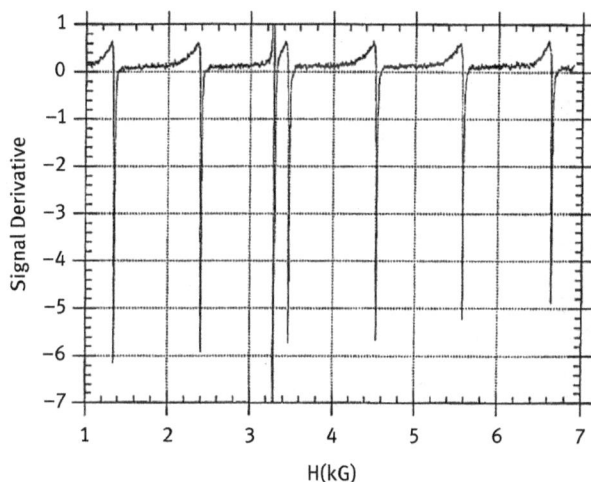

Fig. 4.12: Low temperature ($T = 4.2$ K) X-band ($v = 9.26$ GHz) EPR spectrum of Pr(1%):LaTFMS with $\vec{B}_1 \parallel \vec{B}_0 \parallel c$ – axis. The TFMS compounds have a hexagonal structure with the c-axis along the long dimension of the crystal. This is an integer-spin system with $I = 5/2$ resulting in six highly asymmetric hyperfine lines as shown. The very intense line at ~ 3.3 kG is due to a free radical marker (DPPH). This marker was in the form of a small grain which was placed inside the cavity at a location where $\vec{B}_1 \perp \vec{B}_0$. Such a marker is useful in accurately calibrating the magnetic field. Spectrum taken by the author using a homebuilt X-band EPR spectrometer equipped with a glass immersion cryostat.

4.2.4.3 Interacting Fe ions in cytochrome c_{554} (cyt c_{554})

This is an example of a non-Kramers system that results due to the interaction between two spins. As discussed in Chapter 5, interacting spins can give rise to a net integer spin for the system. For example, consider two electrons each with spin 1/2. Using the rules for the addition of angular momenta from Chapter 2, we can add the two spins to give rise to a total spin $S_T = 0$, 1, resulting in new states for the two spin systems called singlet/triplet states. These states are explored further in Chapter 5 but it is sufficient here to show that an integer spin can result from spin–spin interactions.

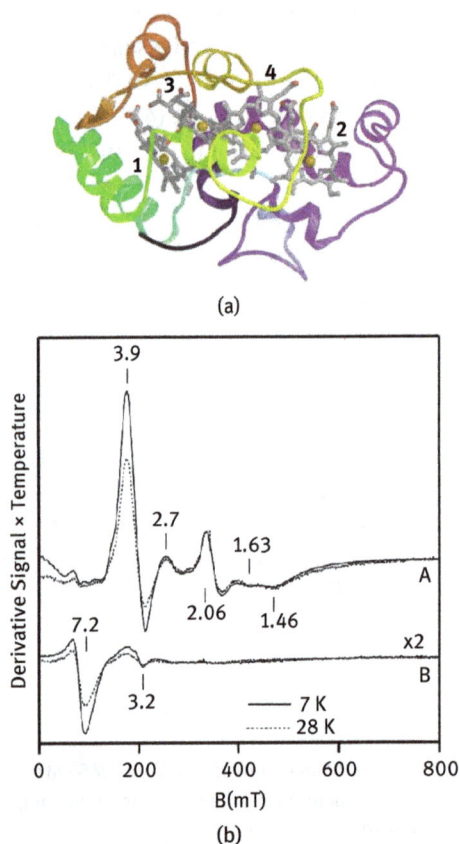

(a)

(b)

Fig. 4.13: Oxidized cytochrome c_{554} is an example of a system that produces EPR signals from both Kramers and non-Kramers species within the same sample. (a) Cartoon structure of the protein showing the four hemes numbered according to their location in the primary sequence [10]. (b) X-band EPR signals from a frozen solution sample taken at two different temperatures as indicated in the figure. Spectrum A was taken with $\vec{B} \perp \vec{B}_1$ and spectrum B with $\vec{B} \parallel \vec{B}_1$. The vertical scale is displayed as a product of the signal and temperature which allows the identification of the Curie law-dependent signals since such signals do not change in intensity at different temperatures [11].

Cyt c_{554} is a tetraheme cytochrome from the chemoautotrophic bacteria *Nitroso-monas europaea* with a molecular weight of 25 kDa that functions as an electron transfer protein in the ammonia oxidation pathway of the biological nitrogen cycle. The protein contains one hs (heme 2) and three ls ferric hemes as shown in Fig. 4.13(a). The hemes are weakly antiferromagnetically coupled in pairs: hemes 2/4 ($S_2 = 5/2$, $S_4 = 1/2$) and hemes 1/3 ($S_1 = 1/2$, $S_3 = 1/2$). No coupling between the pairs was determined. The X-band EPR signals from oxidized cyt c_{554} are shown in Fig. 4.13(b). The signals were obtained from an instrument that employed a bimodal cavity which allowed both polarizations of the microwave magnetic field $\vec{B}_1$ relative to the external magnetic field $\vec{B}$. Signals (A) were obtained with $\vec{B} \perp \vec{B}_1$ (half-integer-spin detection) and in (B) with $\vec{B} \parallel \vec{B}_1$ (integer-spin detection). This metalloprotein is an example of a system with both types of signals. The integer-spin signals were attributed as originating from the coupled ls hemes that result in a singlet/triplet set of states ($S_T = 0, 1$). The half-integer-spin signals originate from the isolated doublets of the hs heme ($S = 5/2$). Figure 4.13(b) also illustrates some of the previous comments in the study of frozen solution samples of metalloproteins. It is clear that the integer-spin signal displayed in spectrum B is highly asymmetric and much weaker than the half-integer-spin signals displayed in spectrum A.

4.3 Crystal field theory

Crystal field theory provides a suitable framework for the understanding of the interactions between a magnetic ion and its crystalline or molecular environment within the context of an introductory exposition to EPR. The amount by which the crystal field splits the energy states of the paramagnetic ion (crystal field splitting) is important in understanding the magnetic properties of the system. The crystal field model represents the ligands as negatively charged points that electrostatically interact with the electrons on the d- and f-orbitals of the metal. No overlap between the ligand valence orbital and metal electron wave functions is allowed in this model making the theory easier to follow. The more involved ligand field theory allows such overlaps where the paramagnetic ion electrons and the ligand electrons occupy molecular orbitals. A discussion of ligand field theory is beyond the scope of this textbook and can be found in other treatises [1, 4, 12].[3] The strength of the crystal field is different in transition series and rare earth ions but the way we treat the crystal field is the same in both cases. This is due to the fact that in the rare earths, the $4f$ electrons are shielded by the outer shells and consequently experience a weaker crystal field. LS coupling is about two orders of magnitude stronger

---

3 In recent years, methods that enhance crystal field and ligand field theories such as ab initio calculations and density functional theory have been developed.

than the crystal field in the rare earths. The situation is reversed in transition series ions where the crystal field is much stronger than LS coupling [13].

The metal electrons experience the electrostatic potential created by the point ligands which obey the Laplace equation:

$$\nabla^2 V_{cf} = 0 \tag{4.14}$$

The solution for the crystal field potential experienced by an electron with coordinates (r, θ, ϕ) and position vector $\vec{r}$ relative to the metal nucleus at the origin of coordinates is given by

$$V(\vec{r}) = \sum_i^N \frac{Ze^2}{|\vec{R}|} = \sum_i^N \frac{Ze^2}{|\vec{r} - \vec{r}_i|} \tag{4.15}$$

where the negatively charged ligands are at position vectors $\vec{r}_i$, and the summation is taken over the total number of ligands. For ligands in an octahedral arrangement, as shown in Fig. 4.14, the summation has six terms ($N = 6$).

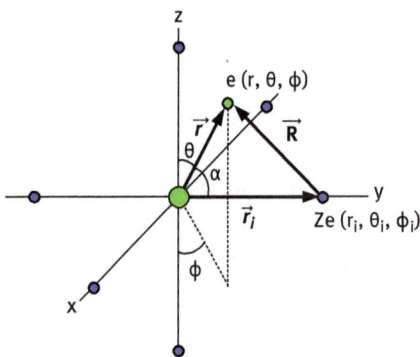

Fig. 4.14: Interaction between a metal electron (small green sphere) with spherical coordinates (r, θ, ϕ) with one of the ligands (blue sphere) from an octahedral arrangement around the metal. The nucleus of the metal is placed at the origin of coordinates (large green sphere). One of the ligands along the y-axis with position vector $\vec{r}_i$ is chosen to define the coordinates of the ligand and the electron.

The $1/|\vec{r} - \vec{r}_i|$ term can be expanded in terms of the Legendre polynomials $P_l(\cos \alpha)$ where α is the angle between the vectors $\vec{r}$ and $\vec{r}_i$ to give [14]

$$\frac{1}{|\vec{r} - \vec{r}_i|} = \sum_{l=0}^{\infty} \frac{r^l P_l(\cos \alpha)}{r_i^{l+1}} \tag{4.16}$$

For cases like the one we are considering here where $r \ll r_i$, the series expansion decreases very quickly and only a few terms are necessary. Summing over all the i ligands allows us to write the crystal field potential as

$$V(\vec{r}) = Ze^2 \sum_i^N \sum_{l=0}^{\infty} \frac{r^l P_l(\cos \alpha_i)}{r_i^{l+1}} \tag{4.17}$$

The addition theorem for spherical harmonics allows the expansion of the Legendre polynomials in terms of spherical harmonics, to reflect the symmetry of the metal orbitals, according to [15]

$$P_l(\cos \alpha_i) = \frac{4\pi}{2l+1} \sum_{m=-l}^{l} Y_l^{m*}(\theta_i, \varphi_i) Y_l^m(\theta, \varphi) \tag{4.18}$$

where $Y_l^m(\theta_i, \varphi_i)$ are spherical harmonics of degree l and azimuthal quantum number m as defined in Chapter 2 (these harmonics are centered at the nucleus of the metal ion which is the same center used for the spherical harmonics that represent the metal electrons [1]). This substitution yields the following expression for the crystal field potential:

$$V(\vec{r}) = Ze^2 \sum_{l=0}^{\infty} \sum_{m=-l}^{l} \frac{4\pi}{2l+1} \sum_i^N \frac{r^l}{r_i^{l+1}} Y_l^{m*}(\theta_i, \varphi_i) Y_l^m(\theta, \varphi) \tag{4.19}$$

According to group theoretical transformation properties of the $Y_l^m(\theta, \varphi)$ harmonics in an octahedral environment, only harmonics up to $l = 4$ are nonzero for d-electrons and $l = 6$ for f-electrons. Further terms in the series expansion can be eliminated based on the symmetry of the ligands. This is further explained later in this section when an example for a rare earth ion is discussed.

The above expression can be rewritten as [16]

$$V_{cf} = \sum_{l=0}^{\infty} \sum_{m=-l}^{l} \sum_i^N A_l^m \langle r^l \rangle Y_l^m(\theta, \varphi) \tag{4.20}$$

where $\langle r^l \rangle$ is the mean value of the lth power of the d- or f-electron radius, and A_l^m are parameters that represent the strength of the crystal field. As mentioned earlier, symmetry arguments can be used to limit the terms of this expansion. Potential functions for various crystal symmetries are tabulated in the book by Low [12].

4.3.1 Transition series ions in square planar and octahedral symmetries

Consider a metal in a D_{4h} symmetry: the four ligands are arranged on a plane with coordinates $\theta = 90°$ and $\phi = 0°$, 90°, 180° and 270° as shown in Fig. 4.15. Certain terms do not appear in this symmetry (such as $c_4^{\pm 1}$ and $c_4^{\pm 3}$) resulting in the simplified potential

$$V = Ze\sqrt{\pi}\left[\frac{-4\,r^2}{\sqrt{5}\,a^3}Y_2^0 + \frac{r^4}{a^5}\left(Y_4^0 + \sqrt{\frac{35}{18}}(Y_4^4 + Y_4^{-4})\right)\right] \qquad (4.21)$$

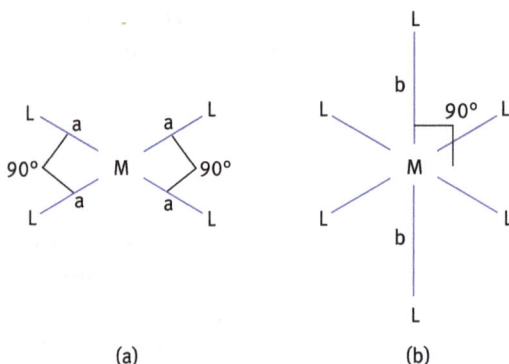

(a)   (b)

Fig. 4.15: Arrangement of ligands in a D_{4h} symmetry: (a) square planar arrangement, (b) six-coordinate tetragonal. M is the metal ion and L is the ligand ion. In pure octahedral symmetry $b = a$.

In an octahedral-axial symmetry where two more ligands are added at a distance b from the metal ion above and below the plane, the coordinates now become $(\theta, \phi) = $ (0, 0), (90, 0), (90, 90), (90, 180), (90, 270) and (180, 0) yielding the crystal field potential

$$V = \sqrt{2\pi}\left\{\begin{array}{l}\sqrt{\frac{49}{18}}\frac{Ze r^4}{a^5}\left[Y_4^0 + \sqrt{\frac{5}{14}}(Y_4^4 + Y_4^{-4})\right] - \\ 2Ze\left[\sqrt{\frac{2}{5}}\left(\frac{r^2}{a^3} - \frac{r^2}{b^3}\right)Y_2^0\right] - 2Ze\left[\sqrt{\frac{2}{9}}\left(\frac{r^4}{a^5} - \frac{r^4}{b^5}\right)Y_4^0\right]\end{array}\right\} \qquad (4.22)$$

which for octahedral symmetry where $b = a$ and the Y_2^0 and Y_4^0 terms disappear to simplify the above expression to

$$V = \frac{Ze r^4}{a^5}\sqrt{\frac{49\pi}{9}}\left[Y_4^0 + \sqrt{\frac{5}{14}}(Y_4^4 + Y_4^{-4})\right] \qquad (4.23)$$

Once the potential functions are determined for a particular symmetry, the crystal field matrix can be constructed with matrix elements given by $\langle \psi_i|V|\psi_j\rangle$ where ψ_i are the general wave functions of all d-orbitals for $m_l = 2, 1, 0, -1, -2$. The resulting 5×5 matrix can be used to write down a secular determinant that can produce the energies of the d-orbitals under the influence of the crystal fields of the various symmetries. The energy states for the pure octahedral and six-coordinate D_{4h} symmetry

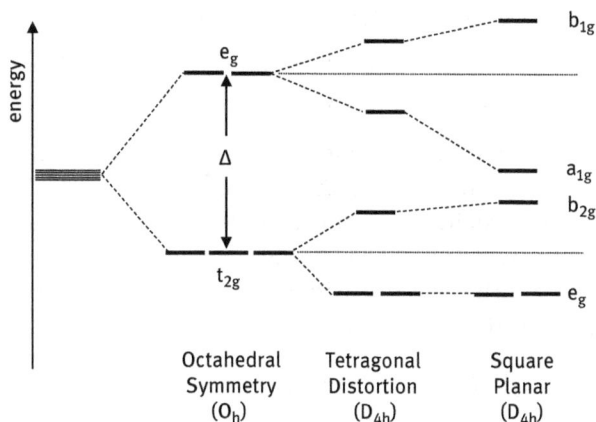

Fig. 4.16: Energy-level diagram of d-orbitals in the presence of O_h and D_{4h} symmetries. The states are labeled according to the group theoretical designations.

were seen earlier in Fig. 4.5. The lowering of the symmetry to the square planar D_{4h} further shifts the tetragonal distortion energy levels as shown in Fig. 4.16.

Detailed calculations for the derivation of the various energy states are found in Lever and Solomon [1].

4.3.2 Rare earth ions in axial symmetry

A straightforward way to manipulate the crystal field Hamiltonian for rare earth ions where there are both orbital and spin contributions to the angular momentum of the electron was introduced by Stevens who replaced the terms in V_{cf} by their equivalent angular momentum operators [17]. This can be done, as long as we stay within a constant J manifold, in the following way: switching to Cartesian coordinates, the operator equivalent of a function $f(x, y, z)$ of a given degree, is found by replacing the coordinates x, y and z by J_x, J_y and J_z, respectively, allowing for the noncommutation of J_x, J_y and J_z. Products of x, y, z are replaced by an expression consisting of all the possible different combinations of J_x, J_y and J_z divided by the total number of combinations. The resulting operator has the same transformation properties under rotation as the potential.

As an example consider a rare earth ion in C_{3h} symmetry where there is a three-fold rotational axis of symmetry with a horizontal plane of reflection. The potential is written as

$$V_{cf}(x_i, y_i, z_i) = \sum_i \left[\begin{array}{l} A_2^0(3z_i^2 - r^2) + A_4^0(35z_i^4 - 30r^2z_i^2 + 3r^4) + \\ A_6^0(231z_i^6 - 315r^2z_i^4 + 105r^4z_i^2 - 5r^6) + \\ A_6^6(x_i^6 - 15x_i^4y_i^2 + 15x_i^2y_i^4 - y_i^6) \end{array} \right] \tag{4.24}$$

where (x_i, y_i, z_i) are the coordinates of a paramagnetic ion electron. The summation is taken over all the $4f$ electrons of the ion. The first three terms are independent of x and y, which means that they are axially symmetric about the z-axis. The fourth term has sixfold symmetry about the z-axis (its form depends on the choice of x-axis relative to the crystal: there are six such choices related by rotations of 60° about z). The equivalent operators O_n^m of the various terms above are defined as

$$\sum (3z^2 - r^2) \equiv \alpha_J \langle r^2 \rangle [3J_z^2 - J(J+1)] \equiv \alpha_J \langle r^2 \rangle O_2^0 \tag{4.25}$$

$$\sum (35z^4 - 30r^2z^2 + 3r^4) \equiv \beta_J \langle r^4 \rangle \left[35J_z^4 - 30J(J+1)J_z^2 + 25J_z^2 - 6J(J+1) + 3J^2(J+1)^2 \right]$$

$$\equiv \beta_J \langle r^4 \rangle O_4^0 \tag{4.26}$$

$$\sum (231z^6 - 315r^2z^4 + 105r^4z^2 - 5r^6)$$

$$\equiv \gamma_J \langle r^6 \rangle [231J_z^6 - 315J(J+1)J_z^4 + 735J_z^4 + 105J^2(J+1)^2J_z^2 - 525J(J+1)J_z^2$$

$$+ 294J_z^2 - 5J^3(J+1)^3 + 40J^2(J+1)^2 - 60J(J+1)] \tag{4.27}$$

$$\equiv \gamma_J \langle r^6 \rangle O_6^0$$

$$\sum (x^6 - 15x^4y^2 + 15x^2y^4 - y^6) \equiv \gamma_J \langle r^6 \rangle \frac{[J_+^6 + J_-^6]}{2} \equiv \gamma_J \langle r^6 \rangle O_6^6 \tag{4.28}$$

These operators are defined within a manifold of given J, and the summation is taken over the coordinates of the magnetic electrons. The numerical coefficients $\alpha_J = \langle J||\alpha||J \rangle$, $\beta_J = \langle J||\beta||J \rangle$ and $\gamma_J = \langle J||\gamma||J \rangle$ are listed in various tables [18]. The crystal field potential now takes the form

$$V_{cf}(x_i, y_i, z_i) = A_2^0 \langle r^2 \rangle \alpha_J O_2^0 + A_4^0 \langle r^4 \rangle \beta_J O_4^0 + A_6^0 \langle r^6 \rangle \gamma_J O_6^0 + A_6^6 \langle r^6 \rangle \gamma_J O_6^6$$

$$= V_2^0 \alpha_J O_2^0 + V_4^0 \beta_J O_4^0 + V_6^0 \gamma_J O_6^0 + V_6^6 \gamma_J O_6^6 \tag{4.29}$$

$$= B_2^0 O_2^0 + B_4^0 O_4^0 + B_6^0 O_6^0 + B_6^6 O_6^6$$

where $V_n^m = A_n^m \langle r^n \rangle$, $B_2^m = V_2^m \alpha_J$, $B_4^m = V_4^m \beta_J$ and $B_6^m = V_6^m \gamma_J$. The B_n^m are now the adjustable parameters to be determined from experiment. For symmetries lower than cubic, the term $B_2^0 O_2^0$ dominates and its effect on the J multiplets is shown in Figs. 4.17 and 4.18 in the case of Kramers and non-Kramers ions.

The potential function reflects the point symmetry of the lattice site; consequently, it must be invariant under the operations of the point group. The less

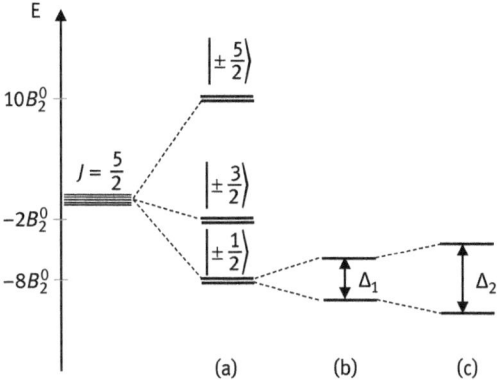

Fig. 4.17: The effect of the term $B_2^0 O_2^0$ on the $J = 5/2$ multiplet of the Ce(III) ion. (a) The degeneracy is lifted to produce three Kramers doublets at energies $-8B_2^0$, $-2B_2^0$ and $10B_2^0$. Splitting of the ground doublet under the influence of an external magnetic field (b) parallel to the axis of symmetry and (c) perpendicular to the axis of symmetry. $\Delta_1 = g_J \beta B$ and $\Delta_2 = 3g_J \beta B$.

symmetric the site, the more potential terms occur in the expansion. All terms containing O_n^m where $l > 6$ are not included for rare earth ions since the matrix elements of V_{cf} in a particular J-manifold vanish. These matrix elements are of the form

$$\int \varphi^*(r_i, \theta_i, \varphi_i) r^l Y_l^m(r_i, \theta_i, \varphi_i) \varphi(r_i, \theta_i, \varphi_i) d\tau \qquad (4.30)$$

where the $\varphi(r_i, \theta_i, \varphi_i)$ are wave functions of the 4f electrons in spherical coordinates. The expansion of the charge density $\varphi^*(r_i, \theta_i, \varphi_i) \cdot \varphi(r_i, \theta_i, \varphi_i)$ in spherical harmonics contains no terms higher than the sixth degree. Therefore, any integral for which $r^l Y_l^m(r_i, \theta_i, \varphi_i)$ is higher than the sixth degree will vanish by the orthogonality relations for spherical harmonics. Terms of odd parity also disappear since $\varphi^*(r_i, \theta_i, \varphi_i) \cdot \varphi(r_i, \theta_i, \varphi_i)$ does not change the sign under parity transformations yielding an integral equal to zero.

The spin operators O_n^0 only connect states with the same $|J_z|$ and so they only contribute diagonal elements to the crystal field matrix. The O_n^0 also contain only even powers of J_z which makes the matrix elements for $-J_z$ the same as the ones for $+J_z$. These terms produce a number of doublets of the type $|\pm J_z\rangle$. The spin operators O_n^m for which $m \neq 0$ have only nondiagonal elements: they admix states of different M to give states of the form $\sum c_M |J, M\rangle$, with the summation taken over M obeying the normalization condition $\sum c_M^2 = 1$. The values of M in the admixture have successive differences equal to m. For example, for the operator O_6^6, the admixed states in a particular J manifold are of the form $c_{M+6}|J, M+6\rangle + c_M|J, M\rangle + c_{M-6}|J, M-6\rangle$. The matrix elements of the operators O_n^m may be obtained from various tables [4]. Once all the matrix elements are calculated, the appropriate matrices for the ion in question

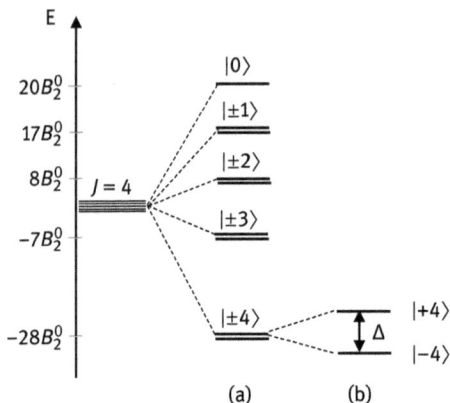

Fig. 4.18: The effect of the term $B_2^0 O_2^0$ on the $J = 4$ multiplets of the Pr(III) ion for the case where $B_2^0 < 0$. (a) A doublet $|\pm 4\rangle$ is the ground state and a singlet $|0\rangle$ is the highest excited state. Jahn–Teller distortions that lower the symmetry can remove the degeneracy of these doublets. (b) Splitting of the ground doublet under the influence of a magnetic field parallel to the axis of symmetry: $\Delta = 8g_J\beta B$.

can be constructed. The problem is then reduced to finding the eigenvalues and eigenfunctions of these matrices and thus determining the crystal field energy levels.

4.3.3 Beyond basic crystal field theory

The preceding discussion demonstrates how, using symmetry arguments, crystal field theory successfully provides a first approximation to the calculation of the energy levels of a paramagnetic ion under the influence of the ligand environment. This is sufficient in some cases but in most systems the energy levels are further shifted relative to the ones predicted by crystal field theory. As already mentioned, this is because crystal field theory does not take into account the overlap between the paramagnetic ion and ligand orbitals and ignores bonding effects since it treats ligands as point charges. These corrections are addressed by the more extensive ligand field theory. The nature of the metal–ligand bond and covalency is included in ligand field theory since ligand orbitals are allowed to overlap and bonding effects take place. The interaction between the metal and ligand orbitals gives rise to metal–ligand bonding and antibonding orbitals whose separation depends on the nature of the ligands, the geometry of the ligands relative to the metal and the distance between metal and ligands. These topics are covered in more advanced textbooks.

References

[1]   A. B. P. Lever and E. I. Solomon, "Ligand Field Theory and the Properties of Transition Metal Complexes," In Inorganic Electronic Structure and Spectroscopy, E. I. Solomon and A. B. P. Lever, Eds., New York, John Wiley & Sons, Inc., 1999, pp. 1–91.

[2]   D. T. Petasis and M. P. Hendrich, "A New Q-Band EPR Probe for Quantitative Studies of Even Electron Metalloproteins," *Journal of Magnetic Resonance*, vol. 136, no. 2, p. 200–206, 1999.

[3]   H. A. Kramers, Théorie générale de la rotation paramagnétique dans les cristaux, *Proceedings of the American Academy of Sciences*, vol. 33, p. 959, 1930.

[4]   A. Abragam and B. Bleaney, Electron Paramagnetic Resonance of Transition Ions, Dover Edition ed., Oxford: Clarendon Press, 1986.

[5]   M. P. Hendrich and P. G. Debrunner, "EPR of Non-Kramers Systems in Biology," In Foundations of Modern EPR, G. R. Eaton, S. S. Eaton and K. M. Salikhov, Eds., Singapore, World Scientific Publishing Co., 1998, pp. 530–547.

[6]   D. T. Petasis and M. P. Hendrich, "Quantitative Interpretation of Multifrequency Multimode EPR Spectra of Metal Containing Proteins, Enzymes, and Biomimetic Complexes," In Methods in Enzymology (vol. 563), P. Z. Qin and K. Warncke, Eds., Amsterdam, Elsevier, 2015, pp. 171–208.

[7]   B. Bleaney and H. E. D. Scovil, "Paramagnetic Resonance in Praseodymium Ethylsulphate," *Philosophical Magazine*, vol. 43, p. 999, 1952.

[8]   J. M. Baker and B. Bleaney, "Hyperfine Structure of Praseodymium," *Proceedings of the Physical Society (London)*, vol. A68, p. 936, 1955.

[9]   D. T. Petasis, Thesis, Pittsburgh: Carnegie Mellon University, 1995.

[10] T. M. Iverson, M. P. Hendrich, D. M. Arciero, A. B. Hopper and D. C. Rees, "Cytochrome c554," In Handbook of Metalloproteins, A Messerschmidt, R Huber, T Poulos and K Wieghardt Eds., Chichester, John Wiley & Sons, Ltd., 2001, pp. 136–146.

[11] A. K. Upadhyay, D. T. Petasis, D. M. Arciero, A. B. Hooper and M. P. Hendrich, "Spectroscopic Characterization and Assignment of Reduction Potentials in the Tetraheme Protein Cytochrome c554 from Nitrosomonus europaea," *Journal of the American Chemical Society*, vol. 125, 2003, pp. 1738–1747.

[12] W. Low, "Paramagnetic Resonance in Solids," In Solid State Physics, Supplement 2, F Seitz and D Turnbull, Eds., New York and London, Academic Press, 1960.

[13] B. Barbara, D. Gignoux and C. Vettier, Lectures on Modern Magnetism (1st ed.), Berlin: Springer-Verlag, 1988.

[14] M. L. Boas, Mathematical Methods in the Physical Sciences (2nd ed.), John Wiley & Sons, New York, 1983.

[15] G. Arfken, Mathematical Methods for Physicists (3rd ed.), Orlando, FL: Academic Press, 1983.

[16] M. T. Hutchings, "Point-Charge Calculation of Energy Levels of Magnetic Ions in Crystalline Electric Fields," *Solid State Physics*, vol. 16, p. 227, 1964.

[17] K. W. H. Stevens, "Matrix Elements and Operator Equivalents Connected with the Magnetic Properties of Rare Earth Ions," *Proceedings of the Physical Society*, vol. A65, 1952, pp. 209–215.

[18] S. Hufner, Optical Spectra of Transparent Rare Earth Compounds, New York: Academic Press, 1978.

5 The spin Hamiltonian

A paramagnetic metal ion can occupy a number of discrete energy states. These spin states are determined by the intrinsic properties of the ion and by interactions between the ion and entities such as the electric dipole moments of the surrounding atoms or molecules (the ligand field), the external magnetic field $\vec{B}$, nuclei that possess a nonzero magnetic moment and interactions with neighboring paramagnetic metal ions. The spin Hamiltonian provides a compact way of representing the energy states of the metal ion in terms of a relatively small number of parameters that can be experimentally determined. The spin Hamiltonian used in data analysis software such as SpinCount incorporates a number of terms to account for the factors mentioned above to correctly calculate the energy states of the metal ion. Even though there can be a large number of spin Hamiltonian terms that can be incorporated within simulation software, most EPR spectra can be analyzed using a small number of terms which will be discussed in this chapter.

The spin Hamiltonian can be used in either the true spin or in an effective spin formalism to describe the low-lying energy levels of a spin system. The effective spin value is determined from the multiplicity $(2S+1)$ of the state. The basis states $|S, M_S\rangle$, where $M_S = S, S-1, \ldots, -S$, serve as the states for the expansion of the spin wave function

$$|\Psi_n\rangle = \sum_i a_{ni}|M_i\rangle \tag{5.1}$$

The spin Hamiltonian operator $\mathcal{H}$ operates only on the spin states $|M_s\rangle$

$$\mathcal{H}|\Psi_n\rangle = E_n|\Psi_n\rangle \tag{5.2}$$

with $n = 1, 2, 3, \ldots, (2S+1)$. E_n are the energies of the spin states and Ψ_n are the eigenfunctions defined above. In order to calculate the energy states we need to determine the matrix elements $\langle M_j|\mathcal{H}|M_i\rangle$ and diagonalize the resulting matrix.

The complete spin Hamiltonian $(\mathcal{H})$ of a system can be expressed in terms of the crystal field $(\mathcal{H}_{cf})$, Zeeman $(\mathcal{H}_{Zeeman})$ and hyperfine $(\mathcal{H}_{hf})$ Hamiltonians. When more than one paramagnetic species is present in the system, the spin–spin interaction $(\mathcal{H}_{int})$ Hamiltonian is included to yield:

$$\mathcal{H} = \mathcal{H}_{cf} + \mathcal{H}_{Zeeman} + \mathcal{H}_{hf} + \mathcal{H}_{int} \tag{5.3}$$

$$\mathcal{H} = \vec{S} \cdot \hat{D} \cdot \vec{S} + \beta \vec{S} \cdot \hat{g} \cdot \vec{B} + \vec{S} \cdot \hat{A} \cdot \vec{I} + \mathcal{H}_{int} \tag{5.4}$$

The discussion in this chapter will show how the spin Hamiltonian is used to calculate the energies of some simple spin states in a paramagnetic system. The book by Orton does an admirable job in showing many of the details of this type of calculations. I will present some of the details here to allow the reader to get an idea of

https://doi.org/10.1515/9783110417562-005

how these calculations are carried out. For a more complete discussion of the calculations, the reader can consult Orton [1].

5.1 Crystal field and Zeeman Hamiltonians

The interaction of the spin system with the ligands and the external magnetic field is given by the crystal field and Zeeman terms

$$\mathcal{H} = \vec{S} \cdot \hat{D} \cdot \vec{S} + \beta \vec{S} \cdot \hat{g} \cdot \vec{B} \tag{5.5}$$

These terms were introduced in Chapter 3. In this section, we will demonstrate how these terms are used in the spin Hamiltonian to determine the energy levels of a spin system. The crystal field Hamiltonian defines the symmetry axes of the molecule or crystal and becomes relevant for paramagnetic systems where $S > 1/2$ and is responsible for the removal of degeneracies when the spin system experiences no external magnetic field. It is typically stronger than the Zeeman interaction, and its effects are taken first in a perturbation calculation.

Two examples of Kramers systems will be considered in some detail: $S = 1/2$ and $S = 3/2$. The same approach can also be applied to systems with $S = 5/2$ (e.g., high-spin Fe(III) and high-spin Mn(II)). One example of a non-Kramers system will be given, namely, the $S = 1$ system.

5.1.1 $S = 1/2$ ground state

The basis functions are $m_S = |\pm \frac{1}{2}\rangle$. We need to define appropriate coordinate axes and we distinguish three cases depending on the symmetry of the crystal field.

5.1.1.1 Cubic symmetry
The g-tensor is isotropic making the choice of coordinate axes arbitrary. Taking the magnetic field along the z-axis yields the spin Hamiltonian

$$\mathcal{H} = \beta \vec{S} \cdot \hat{g} \cdot \vec{B} = g\beta B S_z \tag{5.6}$$

When this spin Hamiltonian operates on the basis wave functions it yields:

$$\mathcal{H}|\psi_n\rangle = g\beta B S_z \left|+\frac{1}{2}\right\rangle = \frac{1}{2}g\beta B \left|+\frac{1}{2}\right\rangle \tag{5.7}$$

$$\mathcal{H}|\psi_n\rangle = g\beta B S_z \left|-\frac{1}{2}\right\rangle = -\frac{1}{2}g\beta B \left|-\frac{1}{2}\right\rangle \tag{5.8}$$

yielding the matrix elements

$$\left\langle \pm\frac{1}{2}\middle|\mathcal{H}\middle|\pm\frac{1}{2}\right\rangle = \pm\frac{1}{2}g\beta B \tag{5.9}$$

which give us the energies

$$E = \pm\frac{1}{2}g\beta B \tag{5.10}$$

which are the expected energy levels we saw in the previous chapter. Using the matrix approach directly, we can express the spin Hamiltonian as

$$
\begin{array}{cc}
 & \begin{array}{cc} |\frac{1}{2}\rangle & |-\frac{1}{2}\rangle \end{array} \\
\begin{array}{c} |\frac{1}{2}\rangle \\ |-\frac{1}{2}\rangle \end{array} & \begin{pmatrix} \frac{1}{2}g\beta B & 0 \\ 0 & -\frac{1}{2}g\beta B \end{pmatrix}
\end{array} \tag{5.11}
$$

which when diagonalized yields the same energies as above.

5.1.1.2 Axial symmetry

Examples are tetragonal and trigonal geometries. The axis of symmetry is taken to be along the z-axis. In this symmetry, the g-factor is anisotropic with $g_z = g_\parallel$ and $g_x = g_y = g_\perp$. The basis states are $|\pm\frac{1}{2}\rangle$. Assuming that the magnetic field lies in the xz-plane making an angle θ with the z-axis simplifies the Zeeman Hamiltonian

$$\mathcal{H} = \beta B \left(g_\parallel \cos\theta S_z + g_\perp \sin\theta S_x \right) \tag{5.12}$$

Using the Pauli matrices $S_z = \frac{1}{2}\begin{pmatrix} 1 & 0 \\ 0 & -1 \end{pmatrix}$ and $S_x = \frac{1}{2}\begin{pmatrix} 0 & 1 \\ 1 & 0 \end{pmatrix}$ yields the matrix

$$
\begin{array}{cc}
 & \begin{array}{cc} |\frac{1}{2}\rangle & |-\frac{1}{2}\rangle \end{array} \\
\begin{array}{c} |\frac{1}{2}\rangle \\ |-\frac{1}{2}\rangle \end{array} & \begin{pmatrix} \frac{1}{2}g_\parallel\beta B\cos\theta & \frac{1}{2}g_\perp\beta B\sin\theta \\ \frac{1}{2}g_\perp\beta B\sin\theta & -\frac{1}{2}g_\parallel\beta B\cos\theta \end{pmatrix}
\end{array} \tag{5.13}
$$

Diagonalizing yields the energies

$$E = \pm\frac{1}{2}\beta B \left(g_\parallel{}^2\cos^2\theta + g_\perp{}^2\sin^2\theta \right)^{1/2} \tag{5.14}$$

When $\vec{B}$ is parallel to the z-axis, $E = \pm\frac{1}{2}g_\parallel\beta B$; and when $\vec{B}$ is perpendicular to the z-axis, $E = \pm\frac{1}{2}g_\perp\beta B$ yielding the energy levels shown in Fig. 3.19(b). In a similar manner, the energies for the rhombic case can be determined as well to yield the energy states shown in Fig. 3.19(c).

Transition probabilities for $S = 1/2$ in axial symmetry: as we have already seen, the transition probability between states n and m is proportional to $\left| \langle \psi_m | \vec{B}_1 \cdot \hat{g} \cdot \vec{S} | \psi_n \rangle \right|^2$, where $\vec{B}_1$ is the magnetic field component of the microwave radiation. In this case, the eigenfunctions are $|\pm \frac{1}{2}\rangle$ and the matrix elements we need to calculate are

$$\left\langle +\frac{1}{2}\Big|\mathcal{H}_1\Big|-\frac{1}{2}\right\rangle = \left\langle +\frac{1}{2}\Big|\vec{B}_1 \cdot \hat{g} \cdot \vec{S}\Big|-\frac{1}{2}\right\rangle \tag{5.15}$$

yielding the same squared matrix elements for the microwave field oriented either along the x- and y-axes

$$\left|\left\langle +\frac{1}{2}\Big|B_1 g_\perp S_x\Big|-\frac{1}{2}\right\rangle\right|^2 = \left|\left\langle +\frac{1}{2}\Big|B_1 g_\perp S_y\Big|-\frac{1}{2}\right\rangle\right|^2 = \frac{1}{4}B_1^2 g_\perp^2 \tag{5.16}$$

and for $\vec{B}_1$ parallel to the z-axis:

$$\left|\left\langle +\frac{1}{2}\Big|B_1 g_\| S_z\Big|-\frac{1}{2}\right\rangle\right|^2 = 0 \tag{5.17}$$

making it clear that transitions only occur when $\vec{B}_1 \perp \vec{B}$.

5.1.2 S = 3/2 ground state

An example of such a system is Cr(III) in octahedral coordination. The spin states are $m_s = \pm 3/2, \pm 1/2$, which are degenerate in zero applied magnetic field.

5.1.2.1 Cubic symmetry
The degeneracy is not lifted by this highly symmetric crystal field. The g-tensor is isotropic and when a magnetic field B is applied, we can express the Zeeman Hamiltonian in the basis states $|\pm\frac{1}{2}\rangle$ and $|\pm\frac{3}{2}\rangle$ following the same approach as in the previous section for an $S = 1/2$ system. The resulting energies are

$$E = \pm \frac{1}{2}g\beta B, \quad \pm \frac{3}{2}g\beta B \tag{5.18}$$

and the energy levels split linearly as shown in Fig. 3.23.

5.1.2.2 Axial symmetry
In the presence of an axial crystal field along the z-axis, the term

$$\mathcal{H} = D\left[S_z^2 - \frac{S(S+1)}{3}\right] \tag{5.19}$$

is introduced. This is the spin Hamiltonian in the absence of an external magnetic field with $S = 3/2$ and the Pauli matrix for S_z given by the diagonal matrix

$$S_z = \begin{pmatrix} 3/2 & & & \\ & 1/2 & & \\ & & -1/2 & \\ & & & -3/2 \end{pmatrix} \tag{5.20}$$

The term $(S(S+1))/3$ makes the Hamiltonian matrix traceless, and its effect on the energy levels is to shift them both by an equal amount. The eigenvalues of this Hamiltonian are

$$E_{|\pm\frac{3}{2}\rangle} = D \text{ and } E_{|\pm\frac{1}{2}\rangle} = -D \tag{5.21}$$

resulting in a splitting equal to 2D between the two Kramers levels as shown in Fig. 3.23. When an external magnetic field B is applied in the xz-plane at an angle θ relative to the z-axis, the spin Hamiltonian takes the form

$$\mathcal{H} = D\left[S_z^2 - \frac{S(S+1)}{3}\right] + \beta B\left(g_{\parallel}\cos\theta S_z + g_{\perp}\sin\theta S_x\right) \tag{5.22}$$

For the special case when the magnetic field is directed along the z-axis, the second term in the parenthesis disappears and the Hamiltonian yields the level energies

$$E_{|\pm\frac{3}{2}\rangle} = D \pm \frac{3}{2}g_{\parallel}\beta B \text{ and } E_{|\pm\frac{1}{2}\rangle} = -D \pm \frac{1}{2}g_{\parallel}\beta B \tag{5.23}$$

which are linearly split versus the magnetic field and are also plotted in Fig. 3.23.

5.1.2.3 Rhombic symmetry

The inclusion of a rhombic term makes the spin Hamiltonian a little more complicated:

$$\mathcal{H} = D\left[S_z^2 - \frac{S(S+1)}{3}\right] + E\left(S_x^2 - S_y^2\right) + \beta\vec{S}\cdot\mathbf{g}\cdot\vec{B} \tag{5.24}$$

where the rhombic term E couples states that differ by $\Delta m = \pm 2$ due to the fact that

$$S_x^2 - S_y^2 = \frac{1}{2}\left(S_+^2 + S_-^2\right) \tag{5.25}$$

When B is along the z-axis, the Hamiltonian reduces to

$$\mathcal{H} = D\left[S_z^2 - \frac{S(S+1)}{3}\right] + E(S_x^2 - S_y^2) + g_z\beta B S_z \tag{5.26}$$

with the Pauli matrices for S_x and S_y given by

$$S_x = \begin{pmatrix} 0 & \sqrt{3}/2 & 0 & 0 \\ \sqrt{3}/2 & 0 & 1 & 0 \\ 0 & 1 & 0 & \sqrt{3}/2 \\ 0 & 0 & \sqrt{3}/2 & 0 \end{pmatrix} \text{ and } S_y = \frac{i}{2} \begin{pmatrix} 0 & -\sqrt{3} & 0 & 0 \\ \sqrt{3} & 0 & -2 & 0 \\ 0 & +2 & 0 & -\sqrt{3} \\ 0 & 0 & \sqrt{3} & 0 \end{pmatrix} \tag{5.27}$$

yielding the spin Hamiltonian matrix

$$\begin{array}{cccc} \left|\frac{3}{2}\right\rangle & \left|-\frac{1}{2}\right\rangle & \left|\frac{1}{2}\right\rangle & \left|-\frac{3}{2}\right\rangle \end{array}$$

$$\begin{array}{c} \left|\frac{3}{2}\right\rangle \\ \left|-\frac{1}{2}\right\rangle \\ \left|\frac{1}{2}\right\rangle \\ \left|-\frac{3}{2}\right\rangle \end{array} \begin{pmatrix} \frac{3}{2}g_\parallel\beta B + D & \sqrt{3}E & & \\ \sqrt{3}E & -\frac{1}{2}g_\parallel\beta B - D & & \\ & & \frac{1}{2}g_\parallel\beta B - D & \sqrt{3}E \\ & & \sqrt{3}E & \frac{3}{2}g_\parallel\beta B + D \end{pmatrix} \tag{5.28}$$

whose terms were rearranged to bring it in a block-diagonal form. Diagonalizing the matrix yields the eigenvalues

$$E = \frac{1}{2}g_z\beta B \pm \left[(D + g_z\beta B)^2 + 3E^2 \right]^{1/2} \tag{5.29a}$$

$$E = -\frac{1}{2}g_z\beta B \pm \left[(D - g_z\beta B)^2 + 3E^2 \right]^{1/2} \tag{5.29b}$$

We see that since the crystal field is no longer an eigenstate of S_z, the levels now diverge in a nonlinear manner relative to B when B values are low with a zero-field splitting equal to

$$\Delta = 2\left(D^2 + 3E^2\right)^{1/2} \tag{5.30}$$

The energy levels become linear at high B values when the term $g\beta B$ is much larger than D and E. It should be noted that due to the nature of the crystal filed coupling, which connects states differing by $\Delta m = \pm 2$, there will now be admixtures of the pure states yielding the admixed states

$$|\pm A\rangle = \alpha \left| \pm \frac{3}{2} \right\rangle + \beta \left| \mp \frac{1}{2} \right\rangle \tag{5.31a}$$

$$|\pm B\rangle = \gamma \left| \pm \frac{1}{2} \right\rangle + \delta \left| \mp \frac{3}{2} \right\rangle \tag{5.31b}$$

where $\alpha \gg \beta$ and $\gamma \gg \delta$. Intradoublet transitions with $\Delta M_s = \pm 2, \pm 3$ which are normally forbidden are now possible within the $|\pm A\rangle$ level that gives rise to a weak EPR signal. These transitions are possible due to the small admixtures of the $\left|\pm\frac{1}{2}\right\rangle$ state within the $\left|\pm\frac{3}{2}\right\rangle$ state but the signal is weak since the intensity is proportional to β^2.

In an axial crystal field ($E = 0$), the energy levels reduce to the ones calculated in the previous section and are plotted in Fig. 5.1(a).

For different orientations of B relative to the molecular axes, we can avoid determining another matrix and instead use the results we obtained for B parallel to the z-axis and do a cyclic permutation of the subscripts. Consider, for example, the case when B is along the x-axis. The spin Hamiltonian in this case becomes

$$\mathcal{H} = D\left[S_z^2 - \frac{S(S+1)}{3}\right] + E\left(S_x^2 - S_y^2\right) + g_x\beta BS_x \tag{5.32}$$

Doing the cyclic permutation $z\to x$, $x\to y$ and $y\to z$ we get

$$\mathcal{H} = D'\left[S_x^2 - \frac{S(S+1)}{3}\right] + E'\left(S_y^2 - S_z^2\right) + g_x\beta BS_x \tag{5.33}$$

Comparing the last two equations and using

$$S_z^2 = S^2 - S_x^2 - S_y^2 = S(S+1) - S_x^2 - S_y^2 \tag{5.34}$$

to eliminate S_z yields

$$\begin{aligned}
&D\left[\frac{2}{3}S(S+1) - S_x^2 - S_y^2\right] + E\left(S_x^2 - S_y^2\right) = \\
&D'\left[S_x^2 - \frac{S(S+1)}{3}\right] + E'\left[S_y^2 - S(S+1) + S_x^2 + S_y^2\right]
\end{aligned} \tag{5.35}$$

Collecting terms and equating yields

$$E' = -\frac{1}{2}(D+E) \quad \text{and} \quad D' = \frac{1}{2}(3E - D) \tag{5.36}$$

Therefore, to get the energy eigenvalues when B is parallel to the x-axis, we make the following substitutions in the expression for the energy eigenvalues for B parallel to the z-axis: $g_z\to g_x$, $D \to \frac{1}{2}(3E - D)$ and $E \to \frac{1}{2}(D - E)$. In a similar way, when B is parallel to the y-axis, we make the substitutions $g_z\to g_y$, $D \to -\frac{1}{2}(3E - D)$ and $E \to \frac{1}{2}(D - E)$.

Applying the previous discussion to an axial case ($E = 0$), we can calculate the energy levels when the magnetic field is perpendicular to the z-axis to get

$$E = \frac{1}{2}g_\perp\beta B \pm \left[\left(-\frac{D}{2} + g_\perp\beta B\right)^2 + \frac{3}{4}D^2\right]^{1/2} \tag{5.37a}$$

$$E = -\frac{1}{2}g_\perp\beta B \pm \left[\left(-\frac{D}{2} - g_\perp\beta B\right)^2 + \frac{3}{4}D^2\right]^{1/2} \tag{5.37b}$$

These energy levels are plotted in Fig. 5.1(b).

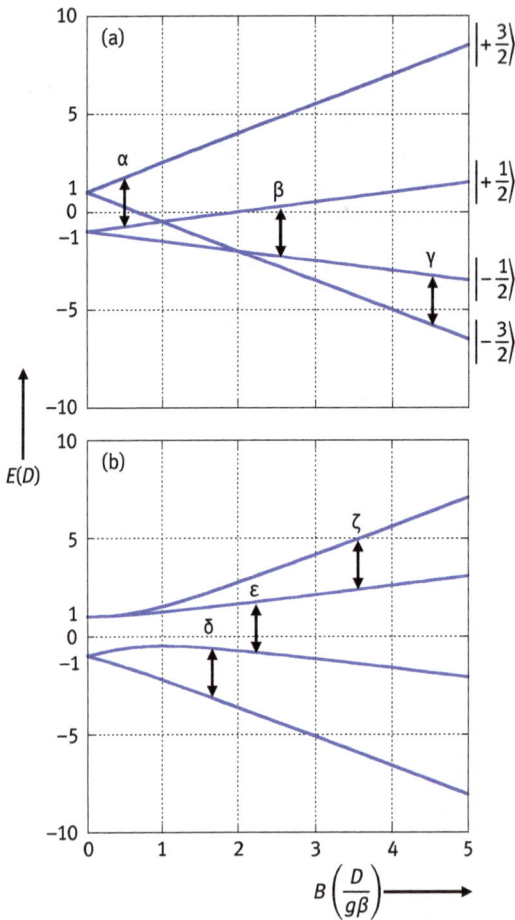

Fig. 5.1: Energy-level splitting for an $S = 3/2$ system experiencing an axial crystal field with $D > 0$ and $D < h\nu$. The magnetic field is parallel to the crystal field axis in (a) and perpendicular in (b). The double arrows show the allowed transitions. When $D < 0$, the order of the doublets is reversed. When $D \gg h\nu$, transition α is not possible and transition γ moves to high field and might not be possible to observe. In the same case in (b), transition ε will not be observed [1].

The doublet separation can be determined from the relative intensities of the EPR lines. The intensity can be easily derived by applying statistical mechanics to a two-level system: for two energy states $|g\rangle$ and $|e\rangle$ with energies E_g and E_e, respectively ($E_e > E_g$), in thermal equilibrium the spin populations are N_g and N_e. Thermal equilibrium occurs when the microwave power is much less than the separation β of the two states so that saturation does not occur. The spin population of a state i is proportional to the Boltzmann factor:

$$N_i \propto e^{-E_i/kT} \tag{5.38}$$

and the intensity of an EPR transition between the two states is proportional to the difference in the populations of the two states:

$$I_{ge} \propto N_g - N_e = e^{-E_g/kT} - e^{-e/kT} \tag{5.39}$$

If the two-level system is part of a general energy manifold of states E_i, we need to account for the thermal population all the energy states leading to the following equation for the intensity of the EPR transition within the two states $|g\rangle$ and $|e\rangle$:

$$I_{ge} \propto N_g - N_e = \frac{e^{-E_g/kT} - e^{-e/kT}}{Z} = \frac{e^{-E_g/kT} - e^{-E_e/kT}}{\sum_i e^{-E_i/kT}} \tag{5.40}$$

where Z is the partition function for the energy manifold and k is the Boltzmann constant. Setting $E_g = 0$ and $\Delta = E_e - E_g$, we get a simplified form of the above equation

$$I_{ge} \propto \frac{1 - e^{-\Delta/kT}}{\sum_i e^{-E_i/kT}} \tag{5.41}$$

This discussion can be applied to transitions α and β in Fig. 5.1, where Δ is equal to $h\nu$ for both, and the ratio of the two intensities becomes simply the ratio of the two partition functions at field positions B_α and B_γ: $I_\gamma/I_\alpha = Z_\alpha/Z_\gamma$. Using the energies

$$E_{\left|\pm\frac{3}{2}\right\rangle} = D \pm \frac{3}{2} g_\parallel \beta B \tag{5.42a}$$

$$E_{\left|\pm\frac{1}{2}\right\rangle} = -D \pm \frac{1}{2} g_\parallel \beta B \tag{5.42b}$$

yields the intensity ratio

$$\frac{I_\gamma}{I_\alpha} = \frac{1 + e^{-\frac{G}{kT}} + e^{\frac{G-2D}{kT}} + e^{\frac{2(G-3D)}{kT}}}{1 + e^{-\frac{G}{kT}} + e^{-\frac{2(G+D)}{kT}} + e^{-\frac{3(G+2D)}{kT}}} \tag{5.43}$$

where $G = g_\parallel \beta B = h\nu$. The value of this intensity ratio approaches one at high temperatures ($kT \gg G$) where both doublets are almost equally populated but has a substantial value at low temperatures. Measurement of the integrated intensities of the two EPR lines at α and γ at a particular temperature T will, consequently, allow the determination of D.

5.1.3 $S = 1$ ground state

As we have already seen, integer-spin systems (non-Kramers) have a different behavior than half-integer systems (Kramers) inside a crystal field. For a system with $S = 1$, the basis states are $|m\rangle = |0\rangle, |\pm1\rangle$. Inside a crystal field of high symmetry such as a cubic field, the states are degenerate and split linearly inside an external magnetic field. Transitions that satisfy the selection rule $\Delta m = \pm1$ are allowed between the $|0\rangle$ and $|\pm1\rangle$ states resulting in two transitions occurring at the same magnetic field value. This can be visualized in Fig. 5.2(a) when $D = 0$ and the two double arrows representing the transitions line up at the same magnetic field value.

5.1.3.1 Axial crystal field

The crystal field term in the Hamiltonian causes a zero-field splitting of the triplet with the $|\pm1\rangle$ state at an energy D above the $|0\rangle$ state when $D > 0$ (the order of the states is reversed when $D < 0$). The application of a magnetic field along the z-axis removes the degeneracy of the $|\pm1\rangle$ state with the individual levels diverging linearly as shown in Fig. 6.2(a). When the magnetic field is oriented perpendicular to the z-axis, the $|m\rangle$ states are no longer pure but include admixtures of the other states. When B is parallel to the x-axis, the triplet states are given by

$$|\psi_0\rangle = \frac{1}{\sqrt{2}}(|+1\rangle - |-1\rangle) \qquad (5.44a)$$

$$|\psi_+\rangle = a(|+1\rangle - |-1\rangle) + b|0\rangle \qquad (5.44b)$$

$$|\psi_-\rangle = a'(|+1\rangle - |-1\rangle) - b'|0\rangle \qquad (5.44c)$$

where a, b, a' and b' are positive constants that depend on $g\beta B$ and D. It is obvious now that there can be transitions between all states as shown in Fig. 6.2(b): the transitions between $|\psi_\pm\rangle$ and $|\psi_0\rangle$ are allowed when the microwave magnetic field B_1 is perpendicular to B and the transition between $|\psi_+\rangle$ and $|\psi_-\rangle$ occurs when B_1 is parallel to B (transition rule $\Delta m = \pm2$) since now matrix elements such as $\langle\psi_-|S_x|\psi_+\rangle$ are no longer zero [1]. The aforementioned transitions are all possible when $D < h\nu$. When, however, $D > h\nu$, the allowed transitions are between the $|0\rangle$ and $|-1\rangle$ states only when $B \parallel z$-axis, and there is now only one transition between the $|\psi_+\rangle$ and $|\psi_0\rangle$ states when $B \perp z$-axis.

5.1.3.2 Rhombic crystal field

A rhombic term $E(S_x^2 - S_y^2)$ needs to be added to the axial crystal field Hamiltonian equation (5.19) like we did in the previous section. This term can also be written as $E(S_+^2 - S_-^2)$ which means that there will be admixtures between the states of the triplet regardless of the orientation of the magnetic field relative to the z-axis. EPR transitions can now occur between all the states of the triplet, even when $B \parallel z$-axis.

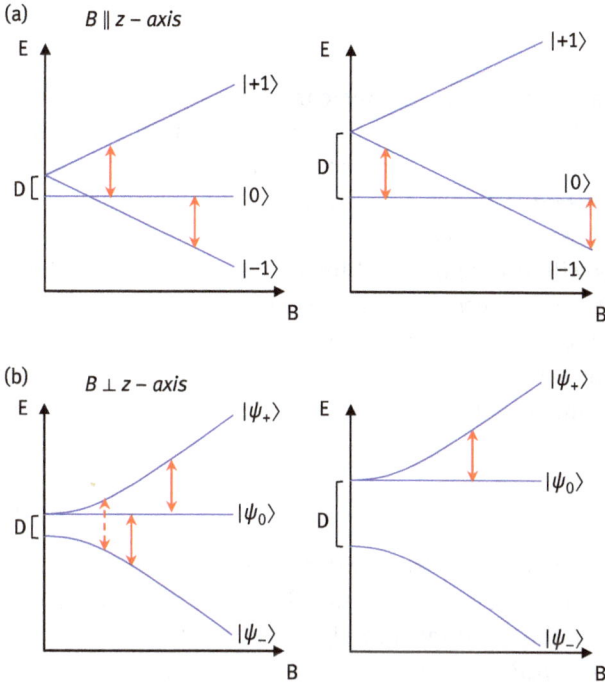

Fig. 5.2: Triplet states of an $S = 1$ system in the presence of an axial crystal field with (a) $B \parallel z$ and (b) $B \perp z$ for the case when $D < h\nu$ (left) and $D > h\nu$ (right). The crystal field removes the degeneracy of the triplet at zero magnetic field. Transitions that obey the selection rule $\Delta m = \pm 1$ are shown by the solid double arrows and occur when $B_1 \perp B$. The dashed arrow represents a normally forbidden transition which is allowed due to admixtures between the states.

5.1.3.3 Half-field transition

Normally forbidden transition that occurs between the $|+1\rangle$ and $|-1\rangle$ states when $B \parallel$ z-axis and $D < h\nu$. This transition occurs at a field that is halfway between zero and $h\nu/$ $g\beta$, the center of the allowed transitions shown in Fig. 5.2(b) on the left. The intensity of the resulting EPR line is weak compared to the intensity of the allowed transitions.

5.2 Hyperfine spin Hamiltonian

As we have already seen in earlier chapters, certain atomic nuclei possess a magnetic dipole moment which quantum mechanically behaves in the same fashion as the magnetic moment of the electron. Because of the existence of the nuclear dipole moment, these nuclei can interact with the magnetic moment of the electron, the external magnetic field and the magnetic field produced by the orbital motion of the electron [2]. A nucleus with nonzero spin I possesses a nuclear moment given by

$$\vec{\mu}_n = \frac{g_n \beta_n}{\hbar} \vec{I} \tag{5.45}$$

where g_n is the nuclear g-factor and β_n is the nuclear magneton. The hyperfine Hamiltonian incorporates all these interactions with $\vec{\mu}_n$ and allows us to study their effects on the EPR spectrum. The interaction between the magnetic moments of the electron and the nucleus is by far the dominant interaction and consequently it will be discussed in more detail. It is represented by the term $\vec{S} \cdot \hat{A} \cdot \vec{I}$ in the spin Hamiltonian. The other interactions will also be outlined first, since they can sometimes have measurable effects on the EPR spectrum. A more complete discussion of these interactions can be found in the treatise by Abragam and Bleaney [3].

The interaction of the nuclear magnetic dipole moment and the magnetic field is called the nuclear Zeeman interaction and in the case of an isotropic interaction is given by

$$\mathcal{H} = -g_n \beta_n \vec{B} \cdot \vec{I} \tag{5.46}$$

This term is typically smaller than the electronic Zeeman and the other hyperfine terms and can usually be ignored. In certain systems, however, this term can be nonnegligible, and its effect on the hyperfine levels is a small shift of the energy of these levels by an amount $\pm \frac{1}{2} g_n \beta_n B$. Because of the negative sign in this term, its effect on the energy levels is opposite to that of the $\vec{S} \cdot \hat{A} \cdot \vec{I}$ term. As a result of this, the energy separation between the hyperfine levels does not change which means that there is no effect on the EPR spectrum produced by the $\vec{S} \cdot \hat{A} \cdot \vec{I}$ term.

Nuclei with $I > 1/2$ have a nonspherical charge distribution which gives rise to an electric quadrupole moment $\vec{\mu}_Q$ which can interact with electric field gradients at the nucleus [4]. The nuclear quadrupole term in the Hamiltonian is given by

$$\mathcal{H} = \vec{I} \cdot \hat{Q} \cdot \vec{I} \tag{5.47}$$

where $\hat{Q}$ is the quadrupole tensor. The effects of the quadrupole interaction on the EPR spectrum are much smaller than the magnetic hyperfine interaction. For a system with axial symmetry about the z-axis, the Hamiltonian takes the form

$$\mathcal{H} = Q \left[I_z^2 - \frac{I}{3}(I+1) \right] \tag{5.48}$$

which leads to a calculation of the effects of the quadrupole interaction on the hyperfine levels. It produces a first-order shift of the energy levels produced by the $\vec{S} \cdot \hat{A} \cdot \vec{I}$ term which is the same for all levels with the same m and, consequently, has no effect on the $\Delta m = 0$ transitions. The second-order terms in addition to the shifts in the energy levels also allow forbidden transitions of the type $\Delta m = \pm 1, \pm 2$ to occur when the magnetic field is rotated away from the z-axis. These transitions produce very weak signals which can be used to detect the presence of quadrupole interaction and measure its strength.

The orbital motion of the electron centered on the nucleus at $r = 0$ produces a magnetic field at the location of the nucleus which can interact with the nuclear magnetic moment. This interaction vanishes when the electron charge density has spherical symmetry [2]. It is, however, observed that there is sometimes a hyperfine interaction when s-electrons are involved due to the fact that, in this case, the probability density of the electronic wave function $|\psi(r = 0)|^2$ has a nonzero value at the position of the nucleus resulting in magnetization

$$\vec{M} = -g\beta|\psi(0)|^2\vec{s} \tag{5.49}$$

experienced by the nucleus. This magnetization gives rise to a term proportional to

$$|\psi(0)|^2\vec{I}\cdot\vec{s} \tag{5.50}$$

in the hyperfine Hamiltonian. This is the Fermi contact term which is important for s-electrons and leads to an isotropic hyperfine interaction with a hyperfine constant given by

$$\frac{8\pi}{3}g\beta g_n\beta_n|\psi(0)|^2 \tag{5.51}$$

The Fermi contact term is zero for d- and f-electrons since their probability density at the nucleus is zero but it is, however, sometimes needed in the analysis of the EPR spectra of d- and f-systems when admixtures of s-orbitals are introduced into d- and f-orbitals [1]. These admixtures are the result of either low symmetry terms or spin polarization where the inner s-levels are polarized by the unpaired spins in outer orbitals resulting in substantial spin density at the nucleus [3].

The interaction between the electronic spin $\vec{S}$ and the nuclear spin $\vec{I}$ dominates the interactions, and in most cases, it is sufficient to use just this interaction to analyze a hyperfine EPR spectrum. It is given by the magnetic hyperfine Hamiltonian defined earlier as

$$\mathcal{H}_{hf} = \vec{S}\cdot\hat{A}\cdot\vec{I} \tag{5.52}$$

where $\hat{A}$ is the hyperfine tensor. In the isotropic case (Fermi contact interaction) where $\hat{A}$ is replaced by a constant A, this Hamiltonian produces in zero field a manifold of states with quantum numbers F ranging from $|S-I|$ to $S+I$ in integer steps and energies given by

$$E_F = \frac{A}{2}[F(F+1) - I(I+1) - S(S+1)] \tag{5.53}$$

For $S = 1/2$, $I = 1/2$, there will be two F states at zero field: a singlet $F = 0$ with energy $E_0 = -3A/4$ and a triplet $F = 1$ with energy $E_1 = A/4$. For $S = 1/2$, $I = 3/2$, the two zero-field states are a triplet $F = 1$ at $E_1 = -5A/4$ and a quintuplet $F = 2$ at $E_2 = 3A/4$. These are shown in Figs. 5.3 and 5.6, respectively.

To apply the Hamiltonian, the nuclear states need to be included along with the electronic states, and the $|m\rangle$ basis states now become $|m, m_I\rangle = |m\rangle|m_I\rangle$ where $m_I = -I$, $-I + 1, \ldots, 0, \ldots, I - 1, I$ are the projections of the nuclear spin I onto the axis of symmetry z. The matrix elements that need to be calculated now are $\langle m, m_I|S_zI_z|m, m_I\rangle$. The operator S_z operates on the $|m\rangle$ states and I_z on the $|m_I\rangle$ states. The same operator rules that apply for electronic states apply for the nuclear states. A couple of examples will be outlined below ($S = 1/2$, $I = 1/2$ and $S = 1/2$, $I = 3/2$ in cubic symmetry and $S = 1/2$, $I = 1/2$ in axial symmetry) to show how the hyperfine Hamiltonian is applied to determine the EPR transitions between hyperfine states. The methodology given here can be extended to systems with different values of S and I and other symmetries [1].

5.2.1 $S = 1/2$ and $I = 1/2$ system in cubic symmetry

In this case, both g and A are isotropic. With the application of a magnetic field parallel to the z-axis, the spin Hamiltonian with the Zeeman and hyperfine terms becomes

$$\mathcal{H} = g\beta BS_z + A\left(S_zI_z + S_xI_x + S_yI_y\right) = g\beta BS_z + A\left[S_zI_z + \frac{1}{2}(S_+I_- + S_-I_+)\right] \quad (5.54)$$

where the spin operators (S_z, S_+ and S_-) act on the electronic parts of the wave function independently of the nuclear spin operators (I_z, I_+ and I_-) which only act on the nuclear parts of the wave function. The basis states are

$$|m, m_I\rangle = \left|\frac{1}{2}, \frac{1}{2}\right\rangle, \left|\frac{1}{2}, -\frac{1}{2}\right\rangle, \left|-\frac{1}{2}, \frac{1}{2}\right\rangle, \left|-\frac{1}{2}, -\frac{1}{2}\right\rangle \quad (5.55)$$

We can operate on these states with $\mathcal{H}$ to get the matrix elements $\langle M, m|\mathcal{H}|M, m\rangle$ that allow us to get the secular determinant and calculate the energy eigenvalues of our spin system:

$|m, m_I\rangle$	$\left|+\frac{1}{2}, +\frac{1}{2}\right\rangle$	$\left|-\frac{1}{2}, -\frac{1}{2}\right\rangle$	$\left|+\frac{1}{2}, -\frac{1}{2}\right\rangle$	$\left|-\frac{1}{2}, +\frac{1}{2}\right\rangle$
$\left|+\frac{1}{2}, +\frac{1}{2}\right\rangle$	$\frac{g\beta B}{2} + \frac{A}{4}$			
$\left|-\frac{1}{2}, -\frac{1}{2}\right\rangle$		$-\frac{g\beta B}{2} + \frac{A}{4}$		
$\left|+\frac{1}{2}, -\frac{1}{2}\right\rangle$			$\frac{g\beta B}{2} - \frac{A}{4}$	$\frac{A}{2}$
$\left|-\frac{1}{2}, +\frac{1}{2}\right\rangle$			$\frac{A}{2}$	$-\frac{g\beta B}{2} - \frac{A}{4}$

$$(5.56)$$

Diagonalizing the block diagonals yields the following eigenvalues [5]:

$$E_1 = \frac{1}{2}g\beta B + \frac{A}{4} \tag{5.57a}$$

$$E_2 = -\frac{1}{2}g\beta B + \frac{A}{4} \tag{5.57b}$$

$$E_3 = \frac{1}{2}\left[(g\beta B)^2 + A^2\right]^{1/2} - \frac{A}{4} = \frac{1}{2}g\beta B\left[1 + \left(\frac{A}{g\beta B}\right)^2\right]^{1/2} - \frac{A}{4} \tag{5.57c}$$

$$E_4 = -\frac{1}{2}\left[(g\beta B)^2 + A^2\right]^{1/2} - \frac{A}{4} = -\frac{1}{2}g\beta B\left[1 + \left(\frac{A}{g\beta B}\right)^2\right]^{1/2} - \frac{A}{4} \tag{5.57d}$$

These energy levels are plotted as a function of $G = g\beta B$ in Fig. 5.3. These energies diverge linearly at high magnetic fields but have a more complicated behavior at low fields. The corresponding wave functions are given by

$$|1\rangle = \left|\frac{1}{2}, \frac{1}{2}\right\rangle \tag{5.58a}$$

$$|2\rangle = \left|-\frac{1}{2}, -\frac{1}{2}\right\rangle \tag{5.58b}$$

$$|3\rangle = \cos\omega\left|\frac{1}{2}, -\frac{1}{2}\right\rangle + \sin\omega\left|-\frac{1}{2}, \frac{1}{2}\right\rangle \tag{5.58c}$$

$$|4\rangle = \cos\omega\left|-\frac{1}{2}, \frac{1}{2}\right\rangle - \sin\omega\left|\frac{1}{2}, -\frac{1}{2}\right\rangle \tag{5.58d}$$

where

$$\sin\omega = \frac{A}{g\beta B} \tag{5.59}$$

The transition probabilities between hyperfine states $|a\rangle$ and $|b\rangle$ when the microwave magnetic field is perpendicular to the z-axis are given by [6]

$$P = |\langle a|g\beta S_x|b\rangle|^2 \tag{5.60}$$

yielding the values

$$P = \left(\frac{1}{2}g\beta\sin\omega\right)^2 \tag{5.61a}$$

for transitions $|1\rangle \rightarrow |3\rangle$ and $|2\rangle \rightarrow |4\rangle$, and

$$P = \left(\frac{1}{2}g\beta\cos\omega\right)^2 \tag{5.61b}$$

for transitions $|1\rangle \rightarrow |4\rangle$ and $|2\rangle \rightarrow |3\rangle$.

The more straightforward picture of $(2I + 1)$ equally spaced hyperfine lines we discussed in Chapter 3 emerges in the limiting case where $g\beta B \gg A$, the limit at which most EPR experiments are carried out. In this case, we can use the binomial theorem to expand the square root in the last two energies to write them as

$$E_3 \approx \frac{1}{2}g\beta B\left[1 + \frac{1}{2}\frac{A^2}{(g\beta B)^2}\right] - \frac{A}{4} = \frac{1}{2}g\beta B - \frac{A}{4} + \frac{1}{4}\frac{A^2}{g\beta B} \tag{5.62a}$$

$$E_4 \approx -\frac{1}{2}g\beta B\left[1 + \frac{1}{2}\frac{A^2}{(g\beta B)^2}\right] - \frac{A}{4} = -\frac{1}{2}g\beta B - \frac{A}{4} + \frac{1}{4}\frac{A^2}{g\beta B} \tag{5.62b}$$

so ignoring the terms in second order in A yields the following energies for the system:

$$E = \pm\frac{1}{2}g\beta B \pm \frac{A}{4} \tag{5.63}$$

which is the same expression we already saw in Chapter 3. At this limit $\omega \approx 0$, resulting in nonzero transition probabilities only for transitions between states $|1\rangle \rightarrow |4\rangle$ and $|2\rangle \rightarrow |3\rangle$ giving rise to two EPR lines of equal intensity separated by A as was shown earlier in Fig. 3.31. Since these transitions are between states $|1\rangle \rightarrow |4\rangle$ and $|2\rangle \rightarrow |3\rangle$, they give rise to the selection rule $\Delta m = \pm 1$ and $\Delta m_I = 0$. It should be noted that in cases where the inequality $g\beta B \gg A$ is not satisfied, the EPR spectrum can be more complicated.

The reason that the nuclear levels for $m = -1/2$ in Fig. 5.3 are inverted relative to those for $m = 1/2$ is due to the fact that the electronic and nuclear magnetic moments have different directions for certain values of m and m_I. For example, the electronic and magnetic moments for $|\frac{1}{2}, -\frac{1}{2}\rangle$ are parallel, whereas for $|\frac{1}{2}, \frac{1}{2}\rangle$ they are in opposite directions making the $|\frac{1}{2}, -\frac{1}{2}\rangle$ level lower in energy. For the $M = -1/2$ levels, the magnetic moments for $|-\frac{1}{2}, \frac{1}{2}\rangle$ are parallel, whereas for $|-\frac{1}{2}, -\frac{1}{2}\rangle$ they are in opposite directions making $|-\frac{1}{2}, \frac{1}{2}\rangle$ lower in energy.

The high-field energy levels are shown in Fig. 5.4, where the allowed transitions are indicated by red double arrows. The energy changes during these transitions are given by

$$\Delta E = E_{|1\rangle} - E_{|4\rangle} = g\beta B_0 + \frac{A}{2} \tag{5.64a}$$

$$\Delta E = E_{|2\rangle} - E_{|3\rangle} = g\beta B_0 - \frac{A}{2} \tag{5.64b}$$

leading to two EPR lines separated by A in constant frequency, field-swept experiment.

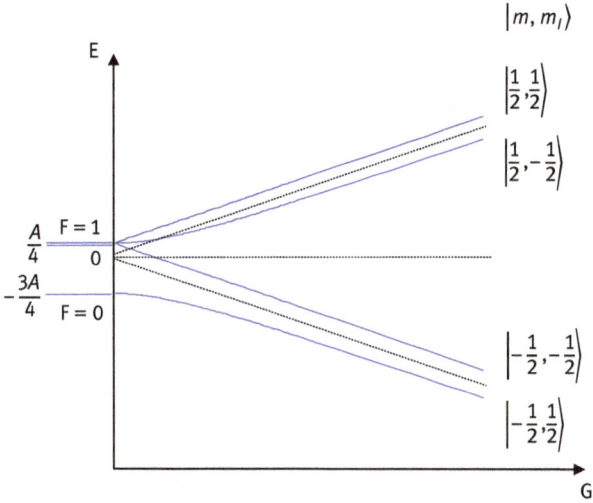

Fig. 5.3: Hyperfine energy levels of a system with $S = 1/2$ and $I = 1/2$ plotted as a function of $G = g\beta B$ (energy axis in units of A). At zero field, we get a triplet state separated by an energy A from a singlet state. At large magnetic fields compared to A, the levels diverge in a linear fashion.

When the nuclear Zeeman term is included, the energy levels become

$$E = \pm \frac{1}{2} g\beta B \pm \frac{A}{4} \mp \frac{1}{2} g_n \beta_n B \tag{5.65}$$

which results in energy levels that are shifted relative to the high-field levels shown in Fig. 5.4. This shift of the energy levels at high field is illustrated in Fig. 5.5. The allowed EPR transitions are shown by the red double arrows. Since the contribution due to the nuclear Zeeman term cancels out, these are the same as in the case where the nuclear Zeeman term is ignored and gives rise to hyperfine lines separated by A.

5.2.2 $S = 1/2$ and $I = 3/2$ system in cubic symmetry

In a similar way, we can determine the energy levels when $S = 1/2$ and $I = 3/2$, where the basis states are now given by

$$|m, m_I\rangle = \left|\frac{1}{2}, \frac{3}{2}\right\rangle, \left|\frac{1}{2}, \frac{1}{2}\right\rangle, \left|\frac{1}{2}, -\frac{1}{2}\right\rangle, \left|\frac{1}{2}, -\frac{3}{2}\right\rangle, \left|-\frac{1}{2}, \frac{3}{2}\right\rangle, \left|-\frac{1}{2}, \frac{1}{2}\right\rangle, \left|-\frac{1}{2}, -\frac{1}{2}\right\rangle, \left|-\frac{1}{2}, -\frac{3}{2}\right\rangle$$

$$\tag{5.66}$$

which yields the following spin Hamiltonian matrix with the terms rearranged in block-diagonal form [1]:

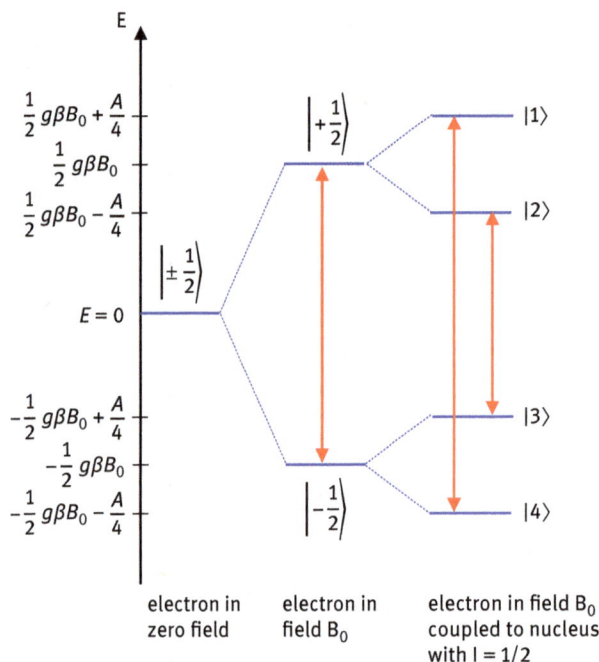

Fig. 5.4: High-field splitting of the hyperfine levels for an $S = 1/2$ and $I = 1/2$ system. In the absence of a hyperfine interaction, only one transition is possible between the electronic levels resulting in a single EPR line. When the electron interacts with a nucleus of spin $I = \frac{1}{2}$, there are two possible transitions giving rise to two hyperfine lines separated by A.

$\|M, m\rangle$	$\left\|\frac{1}{2}, \frac{3}{2}\right\rangle$	$\left\|\frac{1}{2}, \frac{1}{2}\right\rangle$	$\left\|-\frac{1}{2}, \frac{3}{2}\right\rangle$	$\left\|\frac{1}{2}, -\frac{1}{2}\right\rangle$	$\left\|-\frac{1}{2}, \frac{1}{2}\right\rangle$	$\left\|\frac{1}{2}, -\frac{3}{2}\right\rangle$	$\left\|-\frac{1}{2}, -\frac{1}{2}\right\rangle$	$\left\|-\frac{1}{2}, -\frac{3}{2}\right\rangle$
$\left\|\frac{1}{2}, \frac{3}{2}\right\rangle$	$\frac{G}{2} + \frac{3A}{4}$							
$\left\|\frac{1}{2}, \frac{1}{2}\right\rangle$		$\frac{G}{2} + \frac{A}{4}$	$\frac{\sqrt{3}}{2}A$					
$\left\|-\frac{1}{2}, \frac{3}{2}\right\rangle$		$\frac{\sqrt{3}}{2}A$	$-\frac{G}{2} - \frac{3A}{4}$					
$\left\|\frac{1}{2}, -\frac{1}{2}\right\rangle$				$\frac{G}{2} - \frac{A}{4}$	A			
$\left\|-\frac{1}{2}, \frac{1}{2}\right\rangle$				A	$-\frac{G}{2} - \frac{A}{4}$			
$\left\|\frac{1}{2}, -\frac{3}{2}\right\rangle$						$\frac{G}{2} - \frac{A}{4}$	$\frac{\sqrt{3}}{2}A$	
$\left\|-\frac{1}{2}, -\frac{1}{2}\right\rangle$						$\frac{\sqrt{3}}{2}A$	$-\frac{G}{2} + \frac{A}{4}$	
$\left\|-\frac{1}{2}, -\frac{3}{2}\right\rangle$								$-\frac{G}{2} + \frac{3A}{4}$

$$(5.67)$$

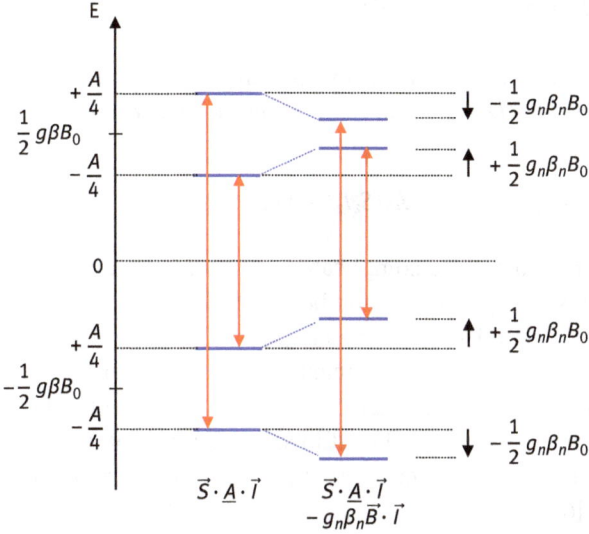

Fig. 5.5: Energy shift of the high-field hyperfine levels for an $S = 1/2$ and $I = 1/2$ system when the magnetic field is set at the value B_0. The double arrows indicate the allowed transitions between the hyperfine levels. It is easy to see that these transitions are not affected by the shift of the energy levels due to the nuclear Zeeman term.

where $G = g\beta B$. Diagonalizing yields the following energies:

$$E_1 = \frac{3}{4}A \pm \frac{G}{2} \tag{5.68a}$$

$$E_2 = -\frac{1}{4}A \pm \frac{\sqrt{G^2 + 2AG + 4A^2}}{2} \tag{5.68b}$$

$$E_3 = -\frac{1}{4}A \pm \frac{\sqrt{G^2 - 2AG + 4A^2}}{2} \tag{5.68c}$$

$$E_4 = -\frac{1}{4}A \pm \frac{\sqrt{G^2 + 4A^2}}{2} \tag{5.68d}$$

These energies are plotted versus G in Fig. 5.6. It is obvious that just as in the $I = 1/2$ example that the energy levels diverge in a linear fashion at high fields but have a more complicated shape at low fields. Following the same analysis as in the $I = 1/2$ example, in the limit when $g\beta B \gg A$ we get four transitions between the hyperfine states resulting in four hyperfine EPR lines of equal intensity separated by A obeying the selection rule $\Delta m = \pm 1$ and $\Delta m_I = 0$ again.

5.2.3 $S = 1/2$ and $I = 1/2$ system in axial symmetry

In the case of axial symmetry, we can have anisotropy in both g and A, and the spin Hamiltonian for the $S = 1/2$ and $I = 1/2$ system in the presence of hyperfine interactions takes the form

$$\mathcal{H} = g_{\parallel}\beta B S_z + g_{\perp}\beta\left(B_x S_x + B_y S_y\right) + A_{\parallel}\left(S_z I_z\right) + A_{\perp}\left(S_x I_x + S_y I_y\right) \tag{5.69}$$

This spin Hamiltonian gives rise to a more complicated energy matrix due to the presence of the S_x and S_y terms that produce a number of off-diagonal terms that makes the matrix difficult to diagonalize. It can be simplified, however, by using perturbation theory where the Zeeman term is assumed to be much larger than the hyperfine term and in the limiting case where $g\beta B \gg A_{\parallel}, A_{\perp}$ to again yield to first-order $(2I+1)$ equally spaced lines. This process is outlined in chapter 3 of Abragam and Bleaney [3], where transitions between states $|m, m_I\rangle \rightarrow |m-1, m_I\rangle$ are found to occur at frequencies given by [6]

$$hv = g\beta B + Am + \frac{A_{\perp}^{2}}{4G}\left(\frac{A_{\parallel}^{2}+A^{2}}{A^{2}}\right)\left[I(I+1)-m^{2}\right]$$

$$+ \frac{\left(A_{\parallel}^{2}-A_{\perp}^{2}\right)^{2}}{2GA^{2}}\left(\frac{g_{\parallel}g_{\perp}}{g^{2}}\right)\left(\sin^{2}\theta\,\cos^{2}\theta\right)m^{2} \tag{5.70}$$

where

$$A^{2} = \frac{A_{\parallel}^{2}\,g_{\parallel}^{2}\cos^{2}\theta + A_{\perp}^{2}\,g_{\perp}^{2}\sin^{2}\theta}{g^{2}} \tag{5.71a}$$

$$g^{2} = g_{\parallel}^{2}\cos^{2}\theta + g_{\perp}^{2}\sin^{2}\theta \tag{5.71b}$$

and θ is the angle between the magnetic field and the z-axis. The $(2I + 1)$ hyperfine lines are separated by A.

We can finish this section by considering the entire hyperfine Hamiltonian and to summarize the effect of the various terms on the EPR spectrum. The complete Hamiltonian is given by

$$\mathcal{H}_{\mathrm{hf}} = \beta\vec{B}\cdot\hat{g}\cdot\vec{S} + \vec{S}\cdot\hat{A}\cdot\vec{I} + \vec{I}\cdot\hat{Q}\cdot\vec{I} - g_n\beta_n\vec{B}\cdot\vec{I} \tag{5.72}$$

with the terms arranged according to their strength. The energy levels for an $S = 1/2$ and $I = 3/2$ system according to this Hamiltonian are shown in Fig. 5.7. The allowed transitions for $\Delta m_I = 0$ are also shown along with the forbidden transitions $\Delta m_I = \pm 1, \pm 2$ which occur under special circumstances. The intensity of the EPR lines due to the forbidden transitions is much smaller than those from allowed transitions.

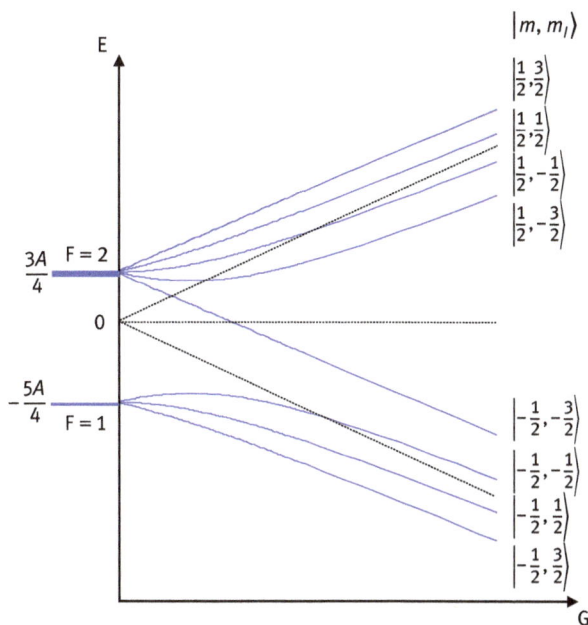

Fig. 5.6: Hyperfine energy levels of a system with $S = 1/2$ and $I = 3/2$ plotted as a function of $G = g\beta B$ (energy axis in units of A). At zero field, a quintuplet state is separated by an energy $2A$ from a singlet state. At large magnetic fields compared to A, the levels diverge in a linear manner.

5.2.4 Interactions with ligand nuclei

This interaction was discussed in Chapter 3 as superhyperfine interaction when an electron on a central ion, in addition to interacting with the magnetic moment of its own nucleus, also interacts with the nuclear magnetic moments of nuclei belonging to neighboring ligands that are coordinated to the central ion. To account for this interaction in the spin Hamiltonian, we simply add up the individual interactions of the electron with each one of the ligand nuclei, which appears as a summation over the interacting nuclei as shown below for an ion in octahedral symmetry

$$\mathcal{H}_{hf} = g\beta \vec{B} \cdot \vec{S} + \sum_i \vec{S} \cdot \hat{A}_i \cdot \vec{I}_i \tag{5.73}$$

In using this Hamiltonian we have to take into account the symmetries experienced by the central ion and the ligands. Consider the paramagnetic ion at the origin of coordinates with the six ligands symmetrically placed along the coordinate axes in an octahedral arrangement as shown in Fig. 4.2. In this case, the central ion experiences a cubic symmetry but the coupling tensors for the interaction with the ligand

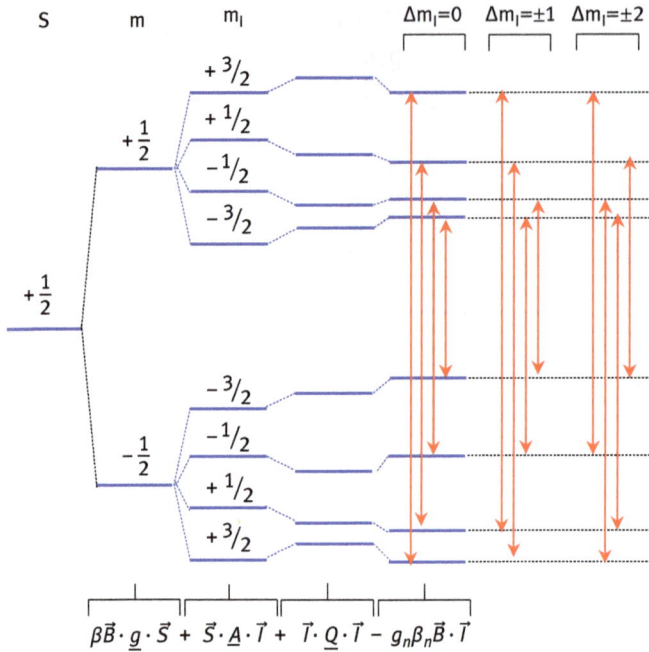

Fig. 5.7: Energy-level diagram for an $S = 1/2$ and $I = 3/2$ system illustrating the effect of various terms of the hyperfine Hamiltonian. The quadrupolar $\vec{I} \cdot .3em-Q \cdot \vec{I}$ and nuclear terms $-g_n\beta_n\vec{B} \cdot \vec{I}$ shift the levels that are split by the $\vec{S} \cdot .3em-A \cdot \vec{I}$ term and allow forbidden transitions to occur which give rise to weak EPR lines (adapted from Fig. 3.13 in Abragam and Bleaney [3]).

nuclei have tetragonal symmetry, with each axis lying along the x-, y-, z-coordinate axes. The resulting energies are given by [2]

$$E = g\beta Bm + \sum_i \frac{\left| \vec{B} \cdot \hat{A}_i \right|}{B} mm_{Ii} \tag{5.74}$$

with $-I_i \le m_{Ii} \le I_i$. This expression holds for nuclei arranged in any symmetry around the central ion. For the octahedral symmetry discussed above the $\hat{A}_i$ tensor has the following values according to the ligand numbering scheme shown in Fig. 5.8:

$$\hat{A}_{1,4} = \begin{pmatrix} A_\parallel & & \\ & A_\perp & \\ & & A_\perp \end{pmatrix} \tag{5.75a}$$

$$\hat{A}_{2,5} = \begin{pmatrix} A_\perp & & \\ & A_\parallel & \\ & & A_\perp \end{pmatrix} \tag{5.75b}$$

$$\hat{A}_{3,6} = \begin{pmatrix} A_\perp & & \\ & A_\perp & \\ & & A_\| \end{pmatrix} \tag{5.75c}$$

and with the magnetic field $\vec{B}$ in an orientation with directional cosines (a, β, γ) relative to the (x, y, z) axes yields the energy values:

$$E = g\beta Bm + \sqrt{A_\|^2\alpha^2 + A_\perp^2(1-\alpha^2)}\,m(m_{I1} + m_{I4})$$
$$+ \sqrt{A_\|^2\beta^2 + A_\perp^2(1-\beta^2)}\,m(m_{I2} + m_{I5}) + \sqrt{A_\|^2\gamma^2 + A_\perp^2(1-\gamma^2)}\,m(m_{I3} + m_{I6}) \tag{5.76}$$

which can be used to find the field positions of the allowed transitions $\Delta m = \pm1$ and $\Delta m_I = 0$:

$$B = \frac{h\nu}{g\beta} - \frac{1}{g\beta}\left[\begin{array}{l}\sqrt{A_\|^2\alpha^2 + A_\perp^2(1-\alpha^2)}(m_{I1} + m_{I4}) + \\ \sqrt{A_\|^2\beta^2 + A_\perp^2(1-\beta^2)}(m_{I2} + m_{I5}) + \sqrt{A_\|^2\gamma^2 + A_\perp^2(1-\gamma^2)}(m_{I3} + m_{I6})\end{array}\right] \tag{5.77}$$

This expression allows us to get the field positions of all the hyperfine lines by allowing each m_{Ii} to take the values $-I_i \le m_{Ii} \le I_i$. We can see that multiple hyperfine lines can occur for the same field: for example, the value $m_{I1} + m_{I4} = 2I - 1$ can occur when either $m_{I1} = I$ and $m_{I4} = I - 1$ or $m_{I1} = I - 1$ and $m_{I4} = I$ result in an EPR line with double the intensity of the line which occurs for $m_{I1} + m_{I4} = 2I$. For example, if we only consider the interaction with two $I = 1/2$ nuclei at ligand positions 1 and 4 with the magnetic field along the x-axis, the above expression becomes

$$B = \frac{h\nu}{g\beta} - \frac{A_\|}{g\beta}(m_{I1} + m_{I4}) \tag{5.78}$$

which yields the following values:

Tab. 5.1: Field positions and relative intensities of hyperfine lines due to interactions with two $I = 1/2$ nuclei.

m_{I1}	m_{I4}	$m_{I1}+m_{I4}$	B	Relative intensity
-1/2	-1/2	-1	$\frac{h\nu}{g\beta} + \frac{A_\|}{g\beta}$	1
1/2	-1/2	0	$\frac{h\nu}{g\beta}$	2
-1/2	1/2	0	$\frac{h\nu}{g\beta}$	
1/2	1/2	+1	$\frac{h\nu}{g\beta} - \frac{A_\|}{g\beta}$	1

According to Tab. 5.1, the interaction of the central spin with the two $I = 1/2$ ligand nuclei 1 and 4 will produce three EPR hyperfine lines with the central line having twice the intensity of the other two lines on either side of the central line. The resulting EPR spectrum is shown in Fig. 3.33. In the same way, we can calculate the interaction with two $I = 1$ nuclei which will produce five symmetrically spaced EPR lines with the intensity ratios 1:2:3:2:1 and with two $I = 3/2$ nuclei there will be seven symmetric lines with the intensity ratios 1:2:3:4:3:2:1 and so on for different values of I. This discussion can be applied to the general case of the interaction with all of the six ligand nuclei in octahedral symmetry where we can start with a single line in the absence of any hyperfine interactions and then add the interaction with ligand nuclei 1 and 4 and then with nuclei 2 and 5 and finally with nuclei 3 and 6 to get the complete spectrum as illustrated in Fig. 5.8.

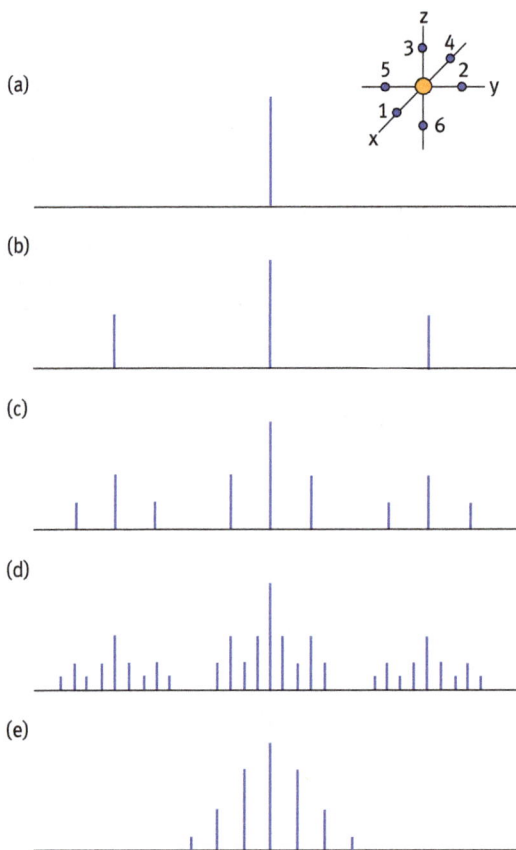

Fig. 5.8: Position and relative intensities of the hyperfine lines for a central spin interacting with octahedrally coordinated equivalent nuclei with $I = 1/2$. (a) No hyperfine interaction, (b) addition of nuclei 1 and 4, (c) addition of nuclei 2 and 5, (d) addition of nuclei 3 and 6 and (e) full spectrum when $\vec{B}$ is directed along the cube diagonal [2].

We should note that due to the symmetry of a system when the field is oriented along certain directions, some of the hyperfine splittings can be equal, causing the number of EPR lines to decrease. For example, in the octahedral symmetry above, when the field is along the cube diagonal, it makes the same angle with all the lines joining the central ion and the six nuclei resulting in some of the hyperfine lines occurring at the same field value resulting in fewer lines as illustrated in Fig. 5.8(e).

The symmetric pattern displayed in Fig. 5.8(e) is representative of the typical spectrum we expect to get when a spin interacts with a number of equivalent nuclei with identical hyperfine splittings for all nuclei. The intensities of the hyperfine lines follow a binomial distribution that was illustrated in Fig. 3.34. In general, if the spin interacts with n equivalent nuclei with spin I, there will be $(2nI + 1)$ hyperfine lines with intensity ratios determined from Pascal's triangle for the appropriate n as illustrated below for $I = 1/2$ and also for $I = 1$.

Tab. 5.2: Pascal's triangle for I = 1/2 and I = 1.

n/I	1/2	1
0	1	1
1	1 1	1 1 1
2	1 2 1	1 2 3 2 1
3	1 3 3 1	1 3 6 7 6 3 1
4	1 4 6 4 1	1 4 10 16 19 16 10 4 1
5	1 5 10 10 5 1	1 5 15 20 45 51 45 20 15 5 1

Applications of this approach were discussed in Section 3.5.1 for the benzene radical anion and the pyrazine radical anion. For benzene, an electron interacts with six equivalent protons each with nuclear spin $I = 1/2$. According to the left triangle in Tab. 5.2, this interaction should result in hyperfine lines with the intensity ratio 1:6:15:20:15:6:1 (by adding the next line of numbers in the triangle) which is observed experimentally as shown in Fig. 3.35(a). In the case of the pyrazine radical anion, there is an interaction with two equivalent nitrogens with $I = 1$, which according to the second triangle in Tab. 5.2 should produce lines with relative intensities 1:2:3:2:1. In addition, there is an interaction with four protons with spin $I = 1/2$ which according to the left triangle in Tab. 5.2 should produce a further splitting with intensity ratio 1:4:6:4:1. The experimental data agrees with this as shown in Fig. 3.35(b). This technique can be extended to interactions with any number of nuclei with any value of I. In reference [2], an analytical expression is quoted for determining the hyperfine line intensities but the direct application of Pascal's triangle method as illustrated above works just as well and is easy to remember.

5.3 Interacting spins

The collection of individual noninteracting paramagnetic ions within a material gives rise to the magnetic properties of the material. Interactions between these ions can modify the magnetic and thermodynamic properties and lead to behavior such as magnetically ordered states below a critical temperature and the appearance of specific heat anomalies at low temperatures in solid-state samples [7]. In fact, such interactions were first discovered by their effect on the magnetic susceptibility, where they caused deviations from the Curie law. In a paramagnetic system (negligible interactions), the Curie law

$$\chi = \frac{C}{T} \tag{5.79}$$

is obeyed where C is the Curie constant and T is the absolute temperature. In the presence of interactions, the Curie–Weiss law

$$\chi = \frac{C}{T+\theta} \tag{5.80}$$

is obeyed, leading to ordering in a ferromagnetic material at the Curie point T_C and in an antiferromagnetic material at the Néel point T_N [8]. The behavior of the susceptibility versus the absolute temperature is illustrated in Fig. 5.9. The bulk properties of a magnetic material as measured by techniques such as magnetic susceptibility, however, can only reveal limited information about the interactions since they measure the average effect of many interionic interactions.

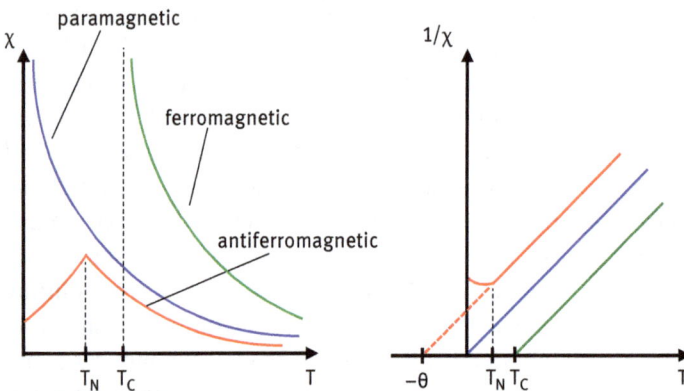

Fig. 5.9: Behavior of the bulk magnetic susceptibility (a) and the inverse susceptibility (b) versus temperature (in Kelvin) of a paramagnetic system and in the presence of ferromagnetic and antiferromagnetic interactions.

It is useful at this point to discuss how the magnetic susceptibility can complement EPR measurements in allowing us to draw a more complete picture of the magnetic properties of a system. In a small applied field B, the nth energy level can be expanded in terms of B according to perturbation theory to yield

$$E_n = E_n^{(0)} + BE_n^{(1)} + B^2 E_n^{(2)} + \cdots \tag{5.81}$$

where $E_n^{(0)}$ is the energy of level n in zero field and $E_n^{(1)}$, $E_n^{(2)}$ are the first- and second-order Zeeman terms which according to perturbation theory can be written in terms of the wave functions $\Psi_n^{(0)}$ of the Hamiltonian in zero field as [9]

$$E_n^{(1)} = \left\langle \Psi_n^{(0)} \middle| \mathcal{H}_{\text{Zeeman}} \middle| \Psi_n^{(0)} \right\rangle \tag{5.82a}$$

$$E_n^{(2)} = \sum_{m \neq n} \frac{\left\langle \Psi_n^{(0)} \middle| \mathcal{H}_{\text{Zeeman}} \middle| \Psi_n^{(0)} \right\rangle^2}{\left(E_n^{(0)} - E_m^{(0)} \right)} \tag{5.82b}$$

This leads to the Van Vleck equation for the magnetic susceptibility for N noninteracting spins at low fields

$$\chi = \frac{N \sum_n \left[\dfrac{\left(E_n^{(1)} \right)^2}{k_B T} - 2 E_n^{(2)} \right] e^{-\frac{E_n^{(0)}}{k_B T}}}{\sum_n e^{-\frac{E_n^{(0)}}{k_B T}}} \tag{5.83}$$

The first term originates from the intrinsic magnetic moment of the particular state and leads to the Curie law. The second term originates from the moment induced by the mixing of low-lying excited states into the ground state by the Zeeman term and is called the Van Vleck temperature-independent paramagnetic term. If the energies E_n are linear in field, then the second term disappears allowing us to simplify the Van Vleck equation by substituting $E_n^{(1)} = \pm \frac{1}{2} g_n \beta B$ to obtain

$$\chi = \frac{N \beta^2}{k_B T} \frac{\sum_n g_n^2 e^{-\frac{E_n^{(0)}}{k_B T}}}{\sum_n e^{-\frac{E_n^{(0)}}{k_B T}}} \tag{5.84}$$

This form of the Van Vleck equation now allows us to correlate information obtained from EPR and χ versus T data on a particular system [10]. For a zero-field energy manifold consisting of Kramers doublets where EPR transitions are allowed within each doublet, the g-values obtained from EPR can be used to fit the χ versus T data and determine the energies of the zero-field states. These values can be compared to estimates of the energies obtained for a variable temperature EPR study of the intensity of the resonance lines originating in each state.

The singlet–triplet states resulting from the interaction of two $S = 1/2$ spins can be thought of as a two-level system. Applying the Van Vleck equation to this system where the singlet and the triplet are separated by an energy J yields an expression that relates the magnetic susceptibility to the exchange interaction constant J [9]:

$$\chi = \frac{2N\beta^2 g^2}{k_B T} \frac{e^{\frac{2J}{k_B T}}}{1 + 3e^{\frac{2J}{k_B T}}} \tag{5.85}$$

This expression holds in the high T or low B limit and allows us to fit the χ versus T data at this limit to obtain an estimate on J.

There are two main heat capacity anomalies in paramagnetic material. The λ anomaly is a sharp feature in the heat capacity that indicates the onset of long-range order in the material. These occur at the Curie temperature for ferromagnetic order and the Néel temperature for antiferromagnetic order. These critical temperatures allow an estimate of the exchange interaction since direct and/or indirect exchange (superexchange) is responsible for long-range order in solids. Let us consider a spin S interacting with n nearest neighbor spins through isotropic exchange that leads to ferromagnetic order at the Curie temperature T_C. In this case, the isotropic exchange and the Curie temperature are related according to [11]

$$J = \frac{3k_B T_C}{nS(S+1)} \tag{5.86}$$

A calorimetric determination of the Curie temperature will, therefore, allow an estimate of J which can be compared to values determined from EPR. The Schottky anomaly, on the other hand, is a rather broad feature in a C versus T plot that is due to the statistical distribution of spins within a two-level system. These systems are readily studied by EPR, and correlating EPR and heat capacity measurements yields a large amount of information about the particular system. For two spin states separated by an energy Δ with a total number of spins equal to N and spin degeneracies N_0 and N_1 for the ground and first excited states, respectively ($N = N_0 + N_1$), the Schottky heat capacity is given by [12]

$$C = Nk_B \left(\frac{\Delta}{k_B T}\right)^2 \frac{e^{\frac{\Delta}{k_B T}}}{\left[1 + e^{\frac{\Delta}{k_B T}}\right]^2} \tag{5.87}$$

A schematic representation of the λ and Schottky anomalies along with the two-level system as defined above is shown in Fig. 5.10. The two states shown in the energy-level diagram can be Kramers doublets that can produce EPR signals as well.

EPR is uniquely qualified to directly measure individual interionic interactions and provide a determination of the nature and strength of interionic interactions. The bulk magnetic properties of the material can then be determined from the information

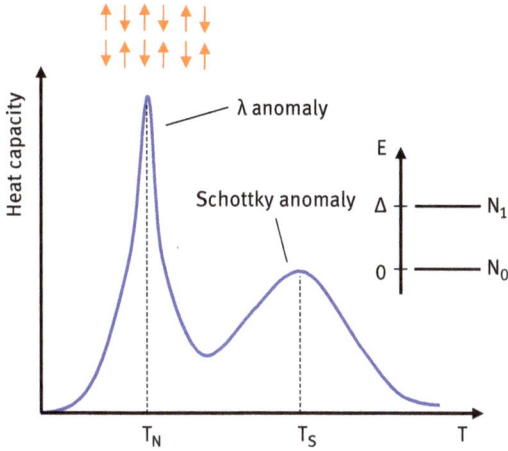

Fig. 5.10: Representation of specific heat anomalies. A sharp λ anomaly indicates the transition to a magnetically ordered state which typically happens at low temperatures for most paramagnetic materials. In the case of antiferromagnetic order, the peak of the λ anomaly occurs at the Néel temperature T_N. A broader more rounded feature at T_S indicates a Schottky anomaly typical of a two-level system.

provided by EPR. The largest contribution to the interactions is due to exchange (direct and superexchange) and magnetic dipole interactions with smaller contributions due to electric quadrupole and interactions through lattice vibrations in crystals (such as virtual phonon exchange in systems with interacting rare earth ions).

5.3.1 Effect of interionic interactions on the EPR spectrum

The interactions act as a perturbation on the energy of an isolated ion, having an effect on the linewidth and position of the EPR lines of the single ion. These effects were first studied by Van Vleck using the theory of moments [13] reaching the following conclusions: at low temperatures when the spin system is nearly aligned by the external magnetic field, the spin–spin interactions (a) shift the position of the resonance line due to the pair of interacting spins relative to that of the single ion since the local field experienced by a particular spin is now perturbed by the field of a neighboring spin, and (b) the increase in the spin alignment leads to a decrease in the spread of field values experienced by a single ion producing a reduction in the linewidth of the EPR line, what we call exchange narrowing. In concentrated samples at low temperatures, the spin–lattice interaction becomes very small and spin–spin interactions are the main source of broadening, dipolar interactions being the largest contributors. The dipolar broadening effects can be very large for concentrated samples, producing linewidths of 10^2–10^3 G which can lead to loss of resolution, depending on the magnitude of the magnetic moments and their separation. Consequently, diluting the paramagnetic species within a diamagnetic host leads to sharper EPR lines. For a particular paramagnetic ion in this case, the chance of another ion occupying a neighboring site is proportional to the fractional concentration. The dominant interaction in this case is between a paramagnetic ion and its nearest neighbor giving rise to a "pair spectrum." For such a case, dilution does not affect the magnitude of the interaction between the two ions. It only reduces the chance of occurrence of such a pair. The single-ion line is due to isolated ions and is consequently much stronger than the pair lines due to the larger abundance of single ions in the sample. As dilution increases, the single-ion linewidth decreases allowing the pair lines to be better resolved. These effects are illustrated in Fig. 5.11.

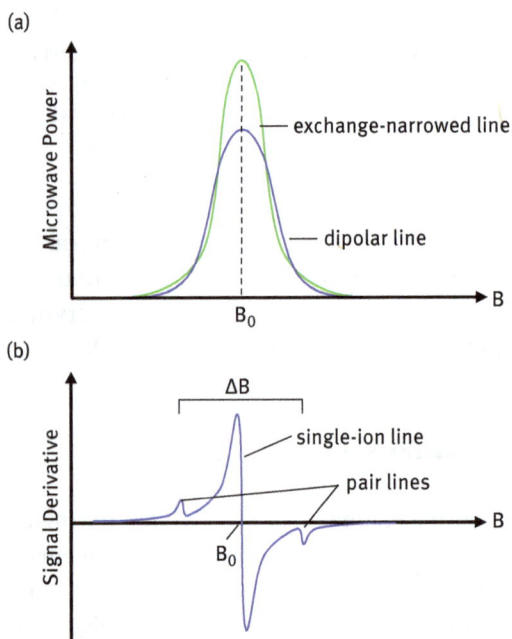

Fig. 5.11: Effect of interionic interactions on the EPR spectrum: (a) when the exchange interaction energy is larger than the dipolar energy, an EPR line broadened by dipolar interactions will experience exchange narrowing while maintaining its Gaussian shape and preserving the area under the absorption line by narrowing in the center part of the line and widening at the wings (for the case when the exchange interaction is much larger than the dipolar energy, the lineshape becomes more Lorentzian in character); (b) in a crystalline sample the interaction between nearest neighbor ions lying on line parallel to the external magnetic field B will produce satellite lines that symmetrically appear on either side of the single-ion line (pair lines). The separation ΔB of the pair lines is a measure of the strength of the interaction between the ions that form the pair.

5.3.2 Nature of interactions

Paramagnetic ions can interact with each other through four types of coupling mechanisms: exchange, magnetic multipole interaction, electric multipole interaction and interaction through lattice vibrations [14, 15]. The last two interactions are typically weaker compared to the first two. The complete Hamiltonian for two interacting spins can be expressed as

$$\mathcal{H} = \mathcal{H}_i + \mathcal{H}_j + \mathcal{H}_{int} \tag{5.88}$$

where the first two terms are the single-ion terms that express the interactions of the single spins with the external magnetic field, crystal field and nuclear moments. The third term contains the above four types of coupling mechanisms that describe the interactions between the two ions.

5.3.2.1 Exchange interaction
Short-range interaction arises from the direct overlap of the electronic wave functions of the two ions (direct exchange) as shown in Fig. 5.12 or through the overlap of the ionic wave functions with an intermediate diamagnetic species (indirect exchange or superexchange).

Direct exchange decreases rapidly with distance between the ions; consequently, it is only important between nearest, and sometimes second nearest, neighbors. It can be thought of in terms of the formation of a weak bond. For example, in the case of the formation of the H_2 molecule, two $S_1 = S_2 = 1/2$ electrons come close together until their wave functions overlap in symmetric and antisymmetric combinations to give rise to a singlet ($S = 0$) and triplet ($S = 1$) total spin states. When a strong bond is formed, the singlet state where the spins are paired up according to the exclusion principle is lowest in energy with the triplet state where the spins are parallel at a much higher energy due to the fact that the probability of this parallel arrangement happening is very small. The triplet state is effectively ignored. In the case of exchange, the interaction between the two spins is much weaker than during bond formation; consequently, the energy separation between singlet and triplet states is much smaller. The triplet state can no longer be ignored since it can be thermally populated at room temperature.

The order of the two states depends upon the nature of the interaction: for an antiferromagnetic interaction, the singlet is lower in energy and for a ferromagnetic interaction, the triplet is the ground state as shown in Fig. 5.13 [16]. In the case of couplings between transition series ions, the interaction can be thought of in terms of a weak bond such as a δ-bond formed between two d-orbitals [17].

For the case when the interacting spins are different than 1/2, they couple to give spin values in the range $|\vec{S}_1 - \vec{S}_2| \leq S \leq \vec{S}_1 + \vec{S}_2$. In the antiferromagnetic case, the total spin state $S = |\vec{S}_1 - \vec{S}_2|$ is the ground state, and in the ferromagnetic case, the $\vec{S} = \vec{S}_1 + \vec{S}_2$ state is the ground state.

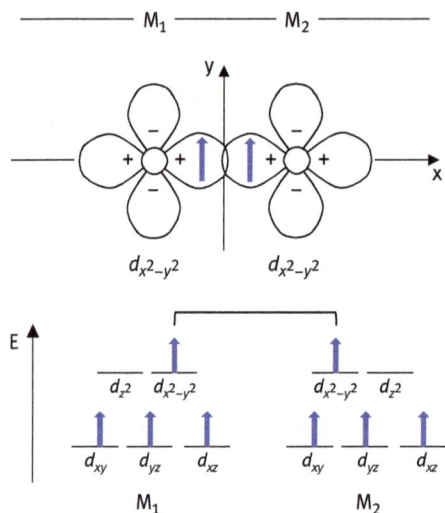

Fig. 5.12: Schematic of a possible direct exchange between electrons on $d_{x^2-y^2}$ orbitals on two metals with $S = 2$ in a linear arrangement. The (+) and (−) signs indicate the sign of the orbital wave functions. The two coupled electrons have the same energy and are on overlapping wave functions. It is evident from this picture that the wave function overlap becomes negligible as the distance between the two metals increases, making exchange interactions short range.

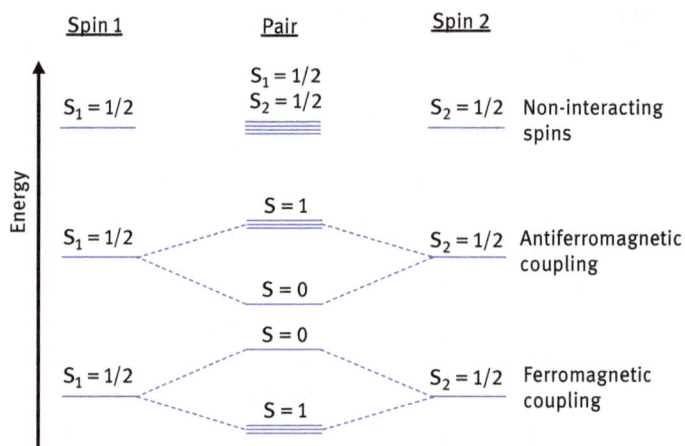

Fig. 5.13: Possible energy levels for a pair of $S_1 = S_2 = 1/2$ interacting spins. Two coupling schemes are possible depending on the order of the singlet and triplet states: in antiferromagnetic coupling, the singlet is the ground state and the triplet is the excited state, whereas in ferromagnetic coupling the state order is reversed.

Indirect exchange occurs via an intermediate diamagnetic ligand. Oxo-bridged metal clusters are found in many systems where the exchange is mediated by the electrons on the p-orbitals of the oxygen. This type of interaction favors antiferromagnetic coupling but ferromagnetic coupling also occurs under certain conditions. Figure 5.14 demonstrates this interaction between two metals with electrons on the same d-orbital in a linear arrangement with the bridging ligand (M_1–L–M_2 angle = 180°). Since the

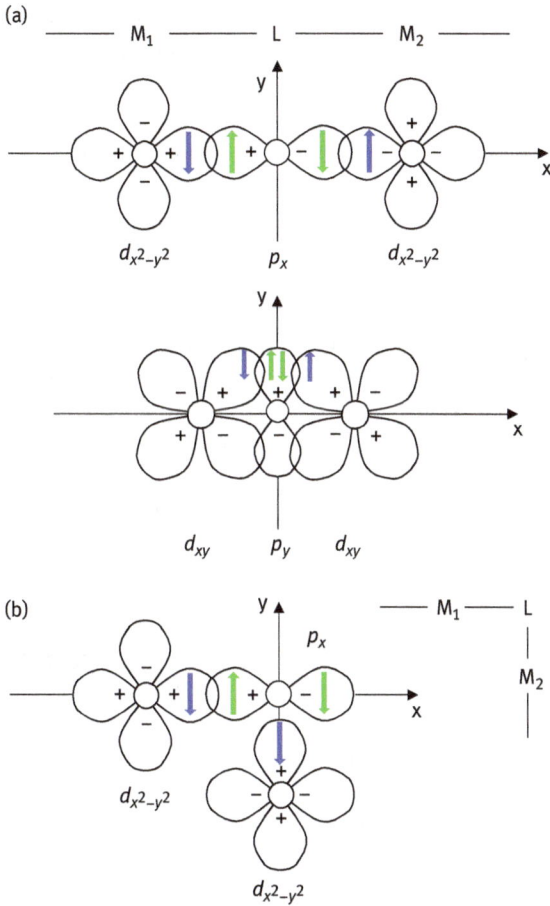

Fig. 5.14: Pictorial representation of two possible mechanisms of indirect exchange between two metals (M_1 and M_2) via a bridging ligand (L) in (a) a linear and (b) an orthogonal arrangement. In the linear arrangement, the exchange occurs due to the overlap of the electronic wave functions of the metal d-orbitals and the ligand p-orbitals. Because the ligand electrons occupy the same orbital, their spins (blue arrows) must be antiparallel producing an antiferromagnetic coupling between the metal spins (green arrows). In the orthogonal arrangement, the overlap between one of the d-orbitals and the p-orbital is zero, allowing the spins on the d-orbitals to remain parallel resulting in a ferromagnetic interaction between the two metals.

electrons on the bridging ligand occupy the same p-orbital, they have to be paired in an antiparallel arrangement according to the exclusion principle. Consequently, it is energetically favored for the electrons on the d-orbitals of the two metal ions to form antiparallel pairs with the ligand electrons, resulting in an antiferromagnetic coupling of the two metal ions. The nature of the interaction depends on which orbitals the metal electrons are on, how many electrons occupy the orbitals and the relative orientation of the orbitals. If the d-orbitals are the same, then the coupling is antiferromagnetic as explained with the strength depending on the relative orientations between the metal d-orbitals and the ligand p-orbitals. When the metal electrons are on different orbitals (e.g., one is on a d_{xy}-orbital and the other on a d_{z^2}-orbital), then ferromagnetic coupling is favored, which is also favored when the metal electrons are on the same orbital but are in an orthogonal instead of a linear arrangement (M_1–L–M_2 angle = 90°) where there is negligible orbital overlap. These are experimental rules proposed by Goodenough and Kanamori and are largely correct [16].

5.3.2.2 Magnetic multipole interaction

The dominant term in the series expansion of the magnetic multipole interactions is the through-space dipolar interaction which is a long-range interaction. This interaction can easily be calculated if the magnetic moment and the position of the ions in the crystal is known. For two interacting dipoles i and j separated by a distance r, the dipolar interaction Hamiltonian is given by

$$\mathcal{H}_{\text{dip}} = \frac{\vec{\mu}_i \cdot \vec{\mu}_j}{r^3} - 3\frac{(\vec{\mu}_i \cdot \vec{r})(\vec{\mu}_j \cdot \vec{r})}{r^5} \tag{5.89}$$

where $\vec{\mu}_i$, $\vec{\mu}_j$ and $\vec{r}$ are as defined in Fig. 5.15. Using the angles θ_i and θ_j of the magnetic moments relative to the position vector $\vec{r}$ allows us to express the Hamiltonian as [18]

$$\mathcal{H}_{\text{dip}} = \frac{\mu_i \mu_j}{r^3} \left(\sin \theta_i \sin \theta_j - 2 \cos \theta_i \cos \theta_j \right) \tag{5.90}$$

The strength of the interaction depends on both the separation between the magnetic moments but also on their orientation relative to the position vector $\vec{r}$ joining the two.

The dipolar Hamiltonian can then be written in terms of the spins using the relation $\vec{\mu}_i = g_i \beta \vec{S}_i$ to yield

$$\mathcal{H}_{\text{dip}} = \frac{g_i g_j \beta^2}{r^3} \left[\vec{S}_i \cdot \vec{S}_j - \frac{3}{r^2} \left(\vec{S}_i \cdot \vec{r} \right) \left(\vec{S}_j \cdot \vec{r} \right) \right] \tag{5.91}$$

For the special case, when the principal axes of the g-tensors and the line joining the ions are parallel to the z-axis, the dipolar Hamiltonian has the form

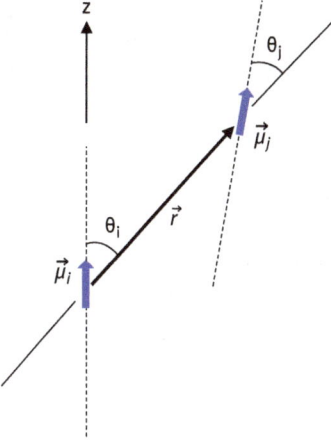

Fig. 5.15: Geometry of two magnetic moments relative to the z-axis separated by a position vector $\vec{r}$. The magnetic dipole–dipole interaction is a long-range through-space interaction and it can couple spins that are separated by several atomic diameters. The external magnetic field B is typically applied parallel to the z-axis.

$$\mathcal{H}_{\text{dip}} = \frac{\beta^2}{r^3} \left(g_{ix}g_{jx}S_{ix}S_{jx} + g_{iy}g_{jy}S_{iy}S_{jy} - 2g_{iz}g_{jz}S_{iz}S_{jz} \right) \tag{5.92}$$

which becomes

$$\mathcal{H}_{\text{dip}} = \frac{\beta^2}{r^3} \left(-2g_{\parallel}^2 S_{iz}S_{jz} + g_{\perp}^2 \left(S_{ix}S_{jx} + S_{iy}S_{jy} \right) \right) \tag{5.93}$$

when the symmetry is axial with the unique axis along the z-axis.

The complete interaction Hamiltonian between spins $\vec{S}_i$ and $\vec{S}_j$ in the presence of both exchange and dipole–dipole interactions is expressed as

$$\mathcal{H}_{\text{int}} = \vec{S}_i \cdot \hat{J}_{ij} \cdot \vec{S}_j + \mathcal{H}_{\text{dip}} \tag{5.94}$$

where $\hat{J}_{ij}$ is the exchange tensor. This tensor can be resolved into an isotropic component J_{ij} and a traceless anisotropic component $\hat{A}_{ij}$ leading to[1]

$$\mathcal{H}_{\text{exch}} = J_{ij}\vec{S}_i \cdot \vec{S}_j + \vec{S}_i \cdot \hat{A}_{ij} \cdot \vec{S}_j \tag{5.95}$$

which is equal to

$$\mathcal{H}_{\text{exch}} = \left(A_{ij}^{\parallel} + J_{ij} \right) S_{iz}S_{jz} + \left(A_{ij}^{\perp} + J_{ij} \right) \left(S_{ix}S_{jx} + S_{iy}S_{jy} \right) \tag{5.96}$$

in the presence of axial symmetry.

---

1 A biquadratic term of the form $-j\left(\vec{S}_i \cdot \vec{S}_j \right)^2$ can also be included but it is very small and ignored in most systems. One of the origins of this term is exchange striction, the phenomenon where exchange causes a decrease of the interionic distance. [16].

For interactions between ions with quenched orbital angular momentum (which includes most transition series ions except heavier ions and Co(II) where spin–orbit coupling cannot be neglected), the anisotropic exchange term is small resulting in the isotropic Heisenberg Hamiltonian [7]

$$\mathcal{H}_{exch} = J_{ij}\vec{S}_i \cdot \vec{S}_j \tag{5.97}$$

The effect of this exchange is the alignment of interacting spins either parallel or antiparallel to each other. Anisotropy in this case is only due to dipolar interactions. In the case of ions with unquenched orbital angular momentum, such as the rare earth ions and some transition series ions, the anisotropic exchange term is not negligible since spin–orbit coupling cannot be ignored.

5.3.2.3 Electric multipole interaction

Electrostatic interaction arises from the nonspherical charge distribution in ions with partially filled electron subshells. For two such interacting ions, the nonspherical charge distribution on each ion can be expanded into a series of electric multipole moments. The interaction between these multipole moments of the two ions leads to a spin–spin interaction [7]. The multipole expansion is that of the Coulomb potential energy term

$$V = \frac{e^2}{r_{ij}} \tag{5.98}$$

which is the only term that contributes to the spin–spin interaction when electron i on one ion interacts with electron j on the other ion with r_{ij} the distance between the two electrons. The dominant term in the expansion is the electric quadrupole interaction [19].

5.3.2.4 Interaction through lattice vibrations

The electric multipole moments of a paramagnetic ion will also interact with the electrostatic crystal field created by the surrounding ligands in the crystal lattice. This interaction is modulated by lattice vibrations which can couple together two neighboring ions. This is due to a magnetic ion participating in the local modes of vibration with nearby ligands which are coupled to lattice vibrations. Ions with fairly large separations can interact through the lattice via the exchange of phonons (*virtual phonon exchange*) making this a long-range interaction [20]. The strength of the interaction depends on the nature of the intervening ligands and, though typically weak, it can be significant in certain systems with unquenched orbital angular momentum [21].

5.3.3 Theory of pair interactions

The case of isotropic exchange between two identical ions (ions experiencing the same crystal field having the same g-values) each with an effective spin $S_{eff} = 1/2$ is the easiest to discuss. As we have already seen, the exchange Hamiltonian is written as

$$\mathcal{H}_{exch} = J_{ij}\vec{S}_i \cdot \vec{S}_j = J\vec{S}_i \cdot \vec{S}_j \qquad (5.99)$$

for simplicity. We should mention here that there are various ways of expressing the exchange Hamiltonian such as $\mathcal{H}_{exch} = -J\vec{S}_i \cdot \vec{S}_j$ or $\mathcal{H}_{exch} = \pm 2J\vec{S}_i \cdot \vec{S}_j$. The particular form of $\mathcal{H}_{exch}$ used will of course have an effect on the actual value of the calculated J. We will be using the form of eq. (5.99) in this discussion where $J > 0$ indicates antiferromagnetic and $J < 0$ ferromagnetic interactions. The total spin $\vec{S}$ that results from the addition of the two spins satisfies the following equation [22]:

$$\vec{S}^2 = \left(\vec{S}_i + \vec{S}_j\right)^2 = S_i^2 + S_j^2 + 2\vec{S}_i \cdot \vec{S}_j \qquad (5.100)$$

where $\vec{S}^2|S\rangle = S(S+1)|S\rangle$. We can now express the product of the two spins as

$$\vec{S}_i \cdot \vec{S}_j = \tfrac{1}{2}\left[S(S+1) - S_i(S_i+1) - S_j(S_j+1)\right] \qquad (5.101)$$

For $\vec{S}_i = \vec{S}_j = \tfrac{1}{2}$ we can then get the energies of the singlet $S = 0$ as $J\vec{S}_i \cdot \vec{S}_j = -3J/4$ and the triplet $S = 1$ as $J\vec{S}_i \cdot \vec{S}_j = J/4$ resulting in the energy-level diagram shown in Fig. 5.16(a).

In the case of isotropic exchange between two identical spins with $S = 1/2$, the levels of the triplet are degenerate in zero magnetic field and lie at an energy equal to J above the singlet state. An applied magnetic field along the z-axis (B_z) will remove the degeneracy of the triplet state but will have no effect on the singlet state resulting in the energy levels ($g_\| = g_z$) shown in Tab. 5.3,

Tab. 5.3: Energy levels due to the isotropic exchange between two identical spins with $S = 1/2$.

S	m_s	Energy
1	+1	$\dfrac{J}{4} + g_\|\beta B_z$
	0	$\dfrac{J}{4}$
	−1	$\dfrac{J}{4} - g_\|\beta B_z$
0	0	$\dfrac{-3J}{4}$

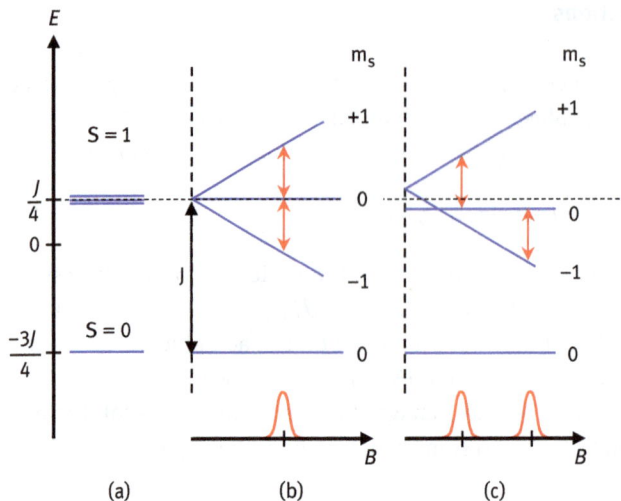

Fig. 5.16: Singlet and triplet states for a pair of ions with $S_{eff} = 1/2$, $J > 0$ and with $B \parallel z$. (a) The $S = 0$ singlet and $S = 1$ triplet are separated by an energy equal to J. (b) For isotropic exchange in the presence of an applied magnetic field, the degeneracy of the m_s levels is removed with the $m_s = \pm 1$ states splitting linearly in B. EPR transitions are allowed between the $m_s = \pm 1$ and $m_s = 0$ states of the triplet as indicated by the arrows resulting in an EPR absorption line, which is at the same magnetic field value as the single-ion line. (c) In the case of anisotropic exchange, the degeneracy of the triplet states is removed in the absence of a magnetic field separating the $m_s = \pm 1$ and $m_s = 0$ levels. The $m_s = \pm 1$ levels further split linearly in B with EPR transitions occurring as in (b). In this case, the transitions do not overlap at the same magnetic field producing two EPR absorption lines as shown. These lines appear as symmetric satellites of the single-ion line in crystalline systems.

which are plotted in Fig. 5.16(b). When anisotropic interactions are present, the triplet is no longer degenerate in zero field but splits into a doublet ($m_s = \pm 1$) and a singlet ($m_s = 0$). The Hamiltonian in this case has a form similar to that for a single spin with $S > 1/2$ as shown in eq. (5.107). This is due to the presence of a traceless anisotropic term which allows us to write the Hamiltonian for the pair of ions interacting along the z-axis as follows [3]:

$$\mathcal{H} = g_z \beta B_z (S_{iz} + S_{jz}) + J_{xx} S_{ix} S_{jx} + J_{yy} S_{iy} S_{jy} + J_{zz} S_{iz} S_{jz} \tag{5.102}$$

Making the substitutions [23]

$$J_{xx} S_{ix} S_{jx} = \frac{1}{2} J_{xx} \left[(S_{ix} + S_{jx})^2 - S_{ix}{}^2 - S_{jx}{}^2 \right] = \frac{1}{2} J_{xx} S_x{}^2 - \frac{1}{4} J_{xx} \tag{5.103a}$$

$$J_{yy} S_{iy} S_{jy} = \frac{1}{2} J_{yy} \left[(S_{iy} + S_{jy})^2 - S_{iy}{}^2 - S_{jy}{}^2 \right] = \frac{1}{2} J_{yy} S_y{}^2 - \frac{1}{4} J_{yy} \tag{5.103b}$$

$$J_{zz}S_{iz}S_{jz} = \frac{1}{2}J_{zz}\left[\left(S_{iz} + S_{jz}\right)^2 - S_{iz}^2 - S_{jz}^2\right] = \frac{1}{2}J_{zz}S_z^2 - \frac{1}{4}J_{zz} \qquad (5.103c)$$

we have substituted the value of each spin as 1/2 to get

$$\mathcal{H} = g_z\beta B_z\left(S_{iz} + S_{jz}\right) + \frac{1}{2}\left(J_{xx}S_x^2 + J_{yy}S_y^2 + J_{zz}S_z^2\right) - \frac{1}{4}\left(J_{xx} + J_{yy} + J_{zz}\right) \qquad (5.104)$$

If we now separate the interaction term into an isotropic part J and a traceless anisotropic term A_{ij} as

$$J_{xx} = J + A_{xx}, \ J_{yy} = J + A_{yy}, \ J_{zz} = J + A_{zz} \qquad (5.105a)$$

where

$$J = \frac{1}{3}\left(J_{xx} + J_{yy} + J_{zz}\right) \qquad (5.105b)$$

$$A_{xx} + A_{yy} + A_{zz} = 0 \qquad (5.105c)$$

After a little manipulation, the Hamiltonian can be expressed as

$$\mathcal{H} = g_\|\beta B_z S_z + \frac{1}{2}J\left[S(S+1) - \frac{3}{2}\right] + \frac{1}{2}\left(A_{xx}S_x^2 + A_{yy}S_y^2 + A_{zz}S_z^2\right) \qquad (5.106)$$

where the third term is the anisotropic exchange term. This results in the pair Hamiltonian

$$\mathcal{H} = g_\|\beta B_z S_z + \frac{1}{2}J\left[S(S+1) - \frac{3}{2}\right] + D\left[S_z^2 - \frac{1}{3}S(S+1)\right] + \frac{1}{2}E\left[S_+^2 + S_-^2\right] \qquad (5.107)$$

The third and fourth terms have the same form as the terms that give rise to the zero-field splittings of single ions as discussed in Section 5.1.3 and illustrated in Fig. 5.2(a) for the case of an ion with $S = 1$. The terms D and E include both exchange and magnetic dipole terms [24]. As we have already seen, the dipolar Hamiltonian has the same form as the anisotropic exchange here. Consequently, experimental determination of D and E using software such as SpinCount [25] allows determination of the anisotropic exchange once the dipolar contribution is subtracted out. The splitting of the triplet depends on the strength of the interaction, with the center of gravity of the energy remaining unchanged. In an applied magnetic field, the levels of the triplet will linearly diverge as shown in Fig. 5.16(c) with transitions only occurring between the $m = \pm 1$ and $m = 0$ levels of the triplet. It is obvious that for isotropic exchange the pair lines will coincide with the single-ion line. For anisotropic exchange, the pair lines appear as satellites of the single-ion line.

Another form of interaction called asymmetric (or antisymmetric) exchange is possible which introduces a term of the form $\hat{d} \cdot (\vec{S}_i \times \vec{S}_j)$ in the exchange Hamiltonian. The physical effect of asymmetric exchange is to create a torque that tends to align the spins at right angles to each other [26]. This term goes to zero for pairs that have a center of inversion symmetry at the middle of the line that joins the two ions and is ignored in highly symmetric cases.

An example will help illustrate the above discussion: consider two similar ions each with $S = 1/2$ in a crystal with uniaxial symmetry and both ions situated along the z-axis. With the magnetic field oriented along the z-axis and in the absence of hyperfine structure, the spin Hamiltonian is

$$\mathcal{H} = \mathcal{H}_i + \mathcal{H}_j + \mathcal{H}_{ij} \tag{5.108}$$

$$
\begin{aligned}
\mathcal{H} &= g_{\|}\beta B_z \left(S_{iz} + S_{jz}\right) + J_{\|}S_{iz}S_{jz} + J_{\perp}\left(S_{ix}S_{jx} + S_{iy}S_{jy}\right) \\
&= g_{\|}\beta B_z \left(S_{iz} + S_{jz}\right) + J_{\|}S_{iz}S_{jz} + \frac{1}{2}J_{\perp}\left(S_{i+}S_{j-} + S_{i-}S_{j+}\right)
\end{aligned} \tag{5.109}
$$

The energy matrix for this spin Hamiltonian in the basis $|S_i, S_j\rangle = \{|++\rangle, |+-\rangle, |-+\rangle, |--\rangle\}$ (where $|++\rangle$ indicates a state where S_i and S_j ($S_{iz} = S_{jz} = 1/2$) are both spin-up, $|+-\rangle$ indicates a state where S_i is spin-up ($S_{iz} = 1/2$) and S_j is spin-down ($S_{jz} = -1/2$), etc.) is

$\lvert S_i, S_j\rangle$	$\lvert++\rangle$	$\lvert+-\rangle$	$\lvert-+\rangle$	$\lvert--\rangle$
$\lvert++\rangle$	$g_{\|}\beta B + \frac{1}{4}J_{\|}$			
$\lvert+-\rangle$		$-\frac{1}{4}J_{\|}$	$\frac{1}{2}J_{\perp}$	
$\lvert-+\rangle$		$\frac{1}{2}J_{\perp}$	$-\frac{1}{4}J_{\|}$	
$\lvert--\rangle$				$-g_{\|}\beta B + \frac{1}{4}J_{\|}$

$$\tag{5.110}$$

The matrix can be block-diagonalized to yield a symmetric triplet and an antisymmetric singlet with the eigenfunctions and eigenvalues given in Tab. 5.4.

Tab. 5.4: Eigenfunctions and eigenvalues of the singlet and triplet states of two $S = 1/2$ interacting spins.

	Eigenfunctions	Eigenvalues
Triplet (symmetric) states	$\lvert++\rangle$	$g_{\|}\beta B + \frac{1}{4}J_{\|}$
	$\frac{1}{\sqrt{2}}(\lvert++\rangle + \lvert-+\rangle)$	$-\frac{1}{4}J_{\|} + \frac{1}{2}J_{\perp}$
	$\lvert--\rangle$	$-g_{\|}\beta B + \frac{1}{4}J_{\|}$
Singlet (antisymmetric) state	$\frac{1}{\sqrt{2}}(\lvert+-\rangle + \lvert-+\rangle)$	$-\frac{1}{4}J_{\|} - \frac{1}{2}J_{\perp}$

Allowed transitions occur when $\Delta S_{iz} = \pm 1$, $\Delta S_{jz} = 0$ or $\Delta S_{iz} = 0$, $\Delta S_{jz} = \pm 1$. Consequently, the only allowed transitions occur within the states of the triplet

$$|++\rangle \leftrightarrow \frac{1}{\sqrt{2}}(|++\rangle + |-+\rangle)$$

with

$$hv = g_\parallel \beta B + \frac{1}{2}(J_\parallel - J_\perp) \tag{5.111a}$$

$$|--\rangle \leftrightarrow \frac{1}{\sqrt{2}}(|++\rangle + |-+\rangle)$$

with

$$hv = g_\parallel \beta B - \frac{1}{2}(J_\parallel - J_\perp) \tag{5.111b}$$

We saw earlier that the exchange tensor can be resolved in terms of an isotropic component J and a traceless anisotropic part A. In an axial symmetry, we can write

$$A_\parallel = \frac{2}{3}(J_\parallel - J_\perp) \tag{5.112}$$

and

$$J = \frac{1}{3}(J_\parallel + 2J_\perp) \tag{5.113}$$

We can now see that with the magnetic field parallel to the z-axis transitions occur when

$$hv = g_\parallel \beta B \pm \frac{1}{2}(J_\parallel - J_\perp) = g_\parallel \beta B \pm \frac{3}{4}A_\parallel \tag{5.114}$$

which demonstrates that, in this special situation, the pair lines appear as symmetric satellites of the single-ion line, and a measurement of the pair line positions yields a measurement of the anisotropic part of the interaction. The anisotropic part of the interaction, in this special case, can therefore be determined directly from the EPR spectrum. The isotropic part of the interaction can also be obtained from a measurement of the intensity of the pair lines as the temperature changes. As shown from Fig. 5.16(a), the separation of the triplet and singlet states is given by J. The population of the triplet states is consequently dependent on J, which can then be measured by comparing the intensity of the EPR lines at two different temperatures.

The case where the interacting ions are the same and have the same spin $S = 1/2$ but they experience slightly different crystal fields also needs to be considered. We have already seen that the g-value is quite sensitive to crystal field effects. Even slight changes in the crystal field experienced by each ion can produce shifts in the g-value. In such a case, the two ions are no longer identical and their pair interactions will yield different results than the ones we obtained above for identical $S = 1/2$ ions. The effects on the pair spectrum depend on how the interaction energy compares to the value of the Zeeman energy $\Delta g \beta B$, where Δg is the difference between the two g-values of the dissimilar ions. For a pair of ions along the z-axis with the magnetic field parallel to the z-axis, the Hamiltonian given by eq. (5.102) now becomes

$$\mathcal{H} = \beta B_z \left(g_{iz} S_{iz} + g_{iz} S_{jz} \right) + J_{xx} S_{ix} S_{jx} + J_{yy} S_{iy} S_{jy} + J_{zz} S_{iz} S_{jz} \tag{5.115}$$

When the Zeeman energy $\Delta g \beta B$ is larger than the interaction energy, diagonalization of $\mathcal{H}$ in the basis $(|++\rangle, |+-\rangle, |-+\rangle, |--\rangle)$ completely removes the degeneracy of the states resulting in the following energies [3]:

$$E_{|++\rangle} = \frac{1}{2} g_z \beta B_z + \frac{1}{4} J_z \tag{5.116a}$$

$$E_{|\pm\rangle} = \frac{1}{2} \Delta g_z \beta B_z - \frac{1}{4} J_z \tag{5.116b}$$

$$E_{|\mp\rangle} = -\frac{1}{2} \Delta g_z \beta B_z - \frac{1}{4} J_z \tag{5.116c}$$

$$E_{|--\rangle} = -\frac{1}{2} g_z \beta B_z + \frac{1}{4} J_z \tag{5.116d}$$

where $g_z = \left(g_{iz} + g_{jz} \right)$ and $\Delta g_z = \left(g_{iz} - g_{jz} \right)$. Transitions are allowed between states $|++\rangle \rightarrow |-+\rangle, |+-\rangle \rightarrow |--\rangle$ and $|++\rangle \rightarrow |+-\rangle, |-+\rangle \rightarrow |--\rangle$ which occur at

$$h\nu = g_{iz} \beta B_z \pm \frac{1}{2} J_z \tag{5.117a}$$

and

$$h\nu = g_{jz} \beta B_z \pm \frac{1}{2} J_z \tag{5.117b}$$

resulting in a spectrum with two pairs of lines, each pair symmetrically placed about the field position where each individual single-ion line appears. The pair spectrum of such ions allows us to determine the value of J_z from the separation of the pair lines instead of just the anisotropic part of the interaction which is the case for identical ions. The values of J_x and J_y can be obtained when the magnetic field is oriented parallel to the x- and y-axes, respectively.

We will now consider two ions just like above but where the difference between the two g-values yields a Zeeman energy $\Delta g \beta B$ of the same order as the interaction strength. The pair spectra now will be different from the ones obtained above. There are now four pair lines placed symmetrically about a single-ion line: two pair lines like the ones shown in Fig. 5.11(b) with an additional pair of satellite lines with a larger magnetic field separation, also centered at the single-ion line. The theory of this type of pair interactions is developed in reference [27].

5.3.4 Pair spectra in the presence of hyperfine interactions

The Hamiltonian for an isolated axial pair of identical ions, i and j, interacting along the z-axis, each having a spin $S = 1/2$ and a nuclear spin I can be written as [28]

$$\mathcal{H} = g_{\parallel} \beta B_z \left(S_{iz} + S_{jz} \right) + A \left(I_{iz} S_{iz} + I_{jz} S_{jz} \right) + J_{\parallel} S_{iz} S_{jz} + J_{\perp} \left(S_{ix} S_{jx} + S_{iy} S_{jy} \right) \tag{5.118}$$

where A is the hyperfine constant. The basis states are a product representation of the effective spin states with the nuclear states $|m_i, m_j\rangle$ where m_i and m_j are the projections of I along the z-axis:

$$|++\rangle = |+1/2, +1/2\rangle |m_i, m_j\rangle \tag{5.119a}$$

$$|+-\rangle = |+1/2, -1/2\rangle |m_i, m_j\rangle \tag{5.119b}$$

$$|-+\rangle = |-1/2, +1/2\rangle |m_i, m_j\rangle \tag{5.119c}$$

$$|--\rangle = |-1/2, -1/2\rangle |m_i, m_j\rangle \tag{5.119d}$$

It is easier to diagonalize the Hamiltonian in the singlet–triplet representation:

$$|1, 1\rangle = |++\rangle \tag{5.120a}$$

$$|1, -1\rangle = |--\rangle \tag{5.120b}$$

$$|1, 0\rangle = \frac{1}{\sqrt{2}} \left(|+-\rangle + |-+\rangle \right) \tag{5.120c}$$

$$|0, 0\rangle = \frac{1}{\sqrt{2}} \left(|+-\rangle - |-+\rangle \right) \tag{5.120d}$$

which yields the following energy matrix with $B \parallel z$-axis:

	$\lvert 1, 1\rangle$	$\lvert 1, -1\rangle$	$\lvert 1, 0\rangle$	$\lvert 0, 0\rangle$
$\lvert 1, 1\rangle$	$g_\parallel \beta B + \dfrac{A}{2} M + \dfrac{1}{4} J_\parallel$			
$\lvert 1, -1\rangle$		$-g_\parallel \beta B - \dfrac{A}{2} M + \dfrac{1}{4} J_\parallel$		
$\lvert 1, 0\rangle$			$\dfrac{2J_{ij}^\perp - J_\parallel}{4}$	$\dfrac{A(\Delta M)}{2}$
$\lvert 0, 0\rangle$			$\dfrac{A(\Delta M)}{2}$	$\dfrac{-2J_{ij}^\perp - J_\parallel}{4}$

$$(5.121)$$

where $M = m_i + m_j$ and $\Delta M = m_i - m_j$. Diagonalization yields the eigenvalues

$$E_1 = g_\parallel \beta B + \frac{A}{2} M + \frac{1}{4} J_\parallel \tag{5.122a}$$

$$E_2 = -g_\parallel \beta B - \frac{A}{2} M + \frac{1}{4} J_\parallel \tag{5.122b}$$

$$E_3 = -\frac{1}{4} J_\parallel + \frac{1}{2} \left[(J_\perp)^2 + A^2 (\Delta M)^2 \right]^{1/2} \tag{5.122c}$$

$$E_4 = -\frac{1}{4} J_\parallel - \frac{1}{2} \left[(J_\perp)^2 + A^2 (\Delta M)^2 \right]^{1/2} \tag{5.122d}$$

EPR transitions within the triplet state satisfy

$$h\nu = g_\parallel \beta B + \frac{A}{2} M \pm \frac{1}{2} J_\parallel \mp \frac{1}{2} \left[(J_\perp)^2 + (A \cdot \Delta M)^2 \right]^{1/2} \tag{5.123}$$

yielding the following expression for the field positions of the resonance lines:

$$B = \frac{h\nu - \frac{AM}{2}}{g_\parallel \beta} + \frac{1}{g_\parallel \beta} \left\{ \mp \frac{1}{2} J_\parallel \pm \frac{1}{2} \left[(J_\perp)^2 + (A \cdot \Delta M)^2 \right]^{1/2} \right\} \tag{5.124}$$

This formula indicates that the lines due to pair interactions (second term) appear as satellites above and below the single-ion hyperfine lines (first term).

It should be noted that the off-diagonal elements produce admixtures between the states $\lvert 0, 0\rangle$ and $\lvert 1, 0\rangle$ which is purely a result of the hyperfine interaction due to the presence of A. The result of this admixture which also requires that the two ions have different z-components of the nuclear spin ($m_i \neq m_j \Rightarrow \Delta M \neq 0$) is that we can now have transitions that are normally forbidden. In the absence of hyperfine interactions and when $m_i = m_j$, only transitions within the components of the triplet can occur, but in the presence of hyperfine interactions, transitions between the triplet and singlet state $\lvert 0, 0\rangle$ which are normally forbidden will also occur. The intensity of these forbidden transitions is proportional to the admixture coefficient and is therefore much weaker than the allowed transitions.

References

[1]   J. W. Orton, Electron Paramagnetic Resonance (An Transition Group Ions in Crystals), New York, Gordon & Breach Science Publishers, Inc., 1969.

[2]   G. E. Pake and T. L. Estle, The Physical Principles of Electron Paramagnetic Resonance, Reading, Massachusetts, W. A. Benjamin, Inc., 1973.

[3]   A. Abragam and B. Bleaney, Electron Paramagnetic Resonance of Transition Ions, Dover Edition ed., Oxford, Clarendon Press, 1986.

[4]   B. Bleaney, "L. Hyperfine structure in paramagnetic salts and nuclear alignment," *Philosophical Magazine*, vol. 42, p. 441–458, 1951.

[5]   P. Rieger, Electron Spin Resonance: Analysis and Interpretation, Cambridge, GBR, Royal Society of Chemistry, 2007.

[6]   B. R. McGarvey, "ESR of Transition Metal Complexes," *Transition Metal Chemistry*, vol. 3, 1967, pp. 88–199.

[7]   J. M. Baker, "Interactions between ions with orbital angular momentum in insulators," *Reports on Progress in Physics*, vol. 34, no. 1, p. 109–173, 1971.

[8]   D. Jiles, Introduction to Magnetism and Magnetic Materials, Boca Raton, FL, CRC Press, 2016.

[9]   O. Kahn, Molecular Magnetism, New York, VCH Publishers, Inc, 1993.

[10]   R. L. Carlin, Magnetochemistry, Berlin-Heidelberg, Springer-Verlag, 1986.

[11]   B. D. Cullity and C. D. Graham, Introduction to Magnetic Materials, New York, John Wiley & Sons, 2011.

[12]   G. Burns, Solid State Physics, Orlando, FL, Academic Press, 1985.

[13]   J. H. Van Vleck, "The dipolar broadening of magnetic resonance lines in crystals," *Physical Review*, vol. 74, no. 9, p. 1168–1183, 1948.

[14]   J. Owen and E. A. Harris, "Pair Spectra and Exchange Interactions," In Electron Paramagnetic Resonance, S. Geschwind, Ed., New York, Plenum Press, 1972, pp. 427–492.

[15]   D. M. S. Bagguley and J. Owen, "Microwave Properties of Solids," *Reports on Progress in Physics*, vol. 20, no. 1, p. 304, 1957.

[16]   A. Bencini and D. Gatteschi, EPR of Exchange Coupled Systems, Heidelberg, Springer-Verlag, 1990.

[17]   F. E. Mabbs and D. J. Machin, Magnetism and Transition Metal Complexes, London, Chapman and Hall, 1973.

[18]   D. J. Griffiths, Introduction to Electrodynamics (4th ed.), Boston, Pearson, 2013.

[19]   W. P. Wolf and R. J. Birgeneau, "Electric Multipole Interactions between Rare Earth Ions," *Physical Review*, vol. 166, p. 376–382, 1968.

[20]   D. H. McMahon and R. H. Silsbee, "Virtual Phonon Effects in the Paramagnetic Resonance of MgO:Fe++," *Physical Review*, vol. 135, no. 1A, p. A91–A96, 1964.

[21]   D. T. Petasis, N. S. VanderVen, S. Simizu and S. A. Friedberg, "Pair interactions in Pr: La $(CF3SO3)3 \cdot 9H2O$," *Journal of Applied Physics*, vol. 73, no. 10, p. 6078–6080, 1993.

[22]   N. W. Ashcroft and N. D. M, Solid State Physics, Holt Rinehart and Wilson, 1976.

[23]   W. Weltner, Magnetic Atoms and Molecules, New York, Dover Publications, 1989.

[24]   S. V. Yurtaeva, A. E. Usachev and Y. V. Yablokov, "Spin–spin interactions of Ce3+ ions in the elpasolite crystal Cs2NaInCl6," *Physics of the Solid State*, vol. 39, no. 7, p. 1075–1080, 1997.

[25] D. T. Petasis and M. P. Hendrich, "Quantitative Interpretation of Multifrequency Multimode EPR Spectra of Metal Containing Proteins, Enzymes, and Biomimetic Complexes," In Methods in Enzymology, vol. *563*, P. Z. Qin and K. Warncke, Eds., Amsterdam, Elsevier, 2015, pp. 171–208.

[26] J. R. Pilbrow, Transition Ion Electron Paramagnetic Resonance, Oxford, Clarendon Press, 1991.

[27] O. Guillot-Noel, P. Goldner, P. Higel and D. Gourier, "A practical analysis of electron paramagnetic resonance spectra of rare earth ion pairs," *Journal of Physics: Condensed Matter*, vol. 16, p. R1–R24, 2004.

[28] J. W. Culvahouse, D. P. Schinke and L. G. Pfortmiller, "Spin-Spin Interaction Constants from the Hyperfine Structure of Pairs of Coupled Ions," *Physical Review*, vol. 177, p. 454–464, 1969.

6 Experimental techniques

Electron paramagnetic resonance (EPR) instrumentation has evolved significantly since the first commercial spectrometers were introduced by Varian more than 50 years ago. Improvements made over the last few decades have enormously improved the sensitivity of EPR spectrometers, and new instrumentation has been developed that allows spectrometers to extend the frequency limits to 100 GHz and beyond, even into the quasi-optical regimes. Most research-grade spectrometers are highly sophisticated pieces of instrumentation with all the microwave plumbing and the electronics hidden behind covers and with computer-controlled automation that give the feeling of a "black box" to the user. This chapter aims to give the reader a basic understanding of the operation of EPR spectrometers for a glimpse of what happens inside the "black box" to allow them to understand how an EPR spectrum is generated.

The fundamental principle of operation of an EPR spectrometer was already introduced in Chapter 1 and illustrated in Fig. 1.12(b) and (c). On top of this basic design, layers of instrumentation must be added to effectively transfer the microwave power to the sample and detector. All standard spectrometers consist of four main components: the microwave bridge, the microwave probe and resonant cavity, an electromagnet and an electronics console. In addition, most spectrometers can be fitted with low-temperature cryostats that allow them to reach temperatures down to around 2 K. These components are shown for an old Varian E-3 spectrometer in Fig. 1.1 and an Elexsys series spectrometer from Bruker in Fig. 1.3. A brief explanation of the function of these components is given further:

Microwave bridge
This contains all the microwave plumbing for the generation, transmission and detection of the microwave radiation.

Microwave probe and resonant cavity
This is the waveguide system that transmits the microwave radiation from the microwave bridge to the resonant cavity, which contains the sample. Modulation coils that allow detection of the reflected signal are usually attached to the walls of the cavity.

Electromagnet
This provides the external magnetic field that gives rise to the Zeeman effect in the sample. Some electromagnets are rotatable to allow different orientations of the magnetic field relative to the sample.

https://doi.org/10.1515/9783110417562-006

Electronics console

This contains all of the electronics instrumentation necessary for the detection and processing of the signal from the detector. In more recent instruments, this console is integrated within the electromagnet power supply.

6.1 Basic low-temperature EPR experiment

Most commercially available EPR spectrometers operate at X-band frequencies (8.5–12 GHz) and employ a continuous-flow liquid helium cryostat to reach very low temperatures (typical temperature range from 2 to 100 K). They utilize a microwave bypass arm to bias a diode detector for increased sensitivity as shown in Fig. 6.1. The radiation from the microwave source (produced by either a klystron or a Gunn oscillator) travels to a resonant cavity where it is stored in the form of standing waves. When resonant absorption of radiation occurs in the sample, a small signal is reflected back from the resonant cavity to the diode detector which acts as a microwave rectifier. The rectified signal from the detector is processed by an electronics console and sent to a computer for recording and analysis. The low temperatures are important to slow electronic relaxation between spin states and to improve the signal-to-noise ratio of the spectrum.

The sample is placed inside the resonant cavity at a position of maximum microwave magnetic field $\vec{B}_1$ as shown in Fig. 6.2. Bimodal rectangular cavities are available that allow the external magnetic field to be oriented in either a perpendicular or a parallel orientation relative to $\vec{B}_1$ for the detection of both half-integer and integer-spin systems, respectively. Alternatively, a cylindrical cavity can be employed with side-wall coupling as in the Q-band (34 GHz) spectrometer in our lab where the external magnetic field can be rotated between the perpendicular and parallel orientations. In spectrometers operating at low temperatures, a continuous-flow cryostat is employed for cooling the sample to liquid helium temperatures. The cold helium gas flows through a nozzle equipped with a heater that allows control of the helium temperature. A thermocouple placed a few millimeters below the sample tube measures the temperature of the gas.

6.2 Main spectrometer components

The sensitivity of an EPR spectrometer is one of the most important criteria required in collecting accurate and reliable spectra. Consequently, a variety of specialized instrumentation is required to achieve this result.

The EPR signal depends on the unpaired electrons in a sample and consequently depends on the magnetic susceptibility of the sample. When the sample is interacting

Microwave Bridge

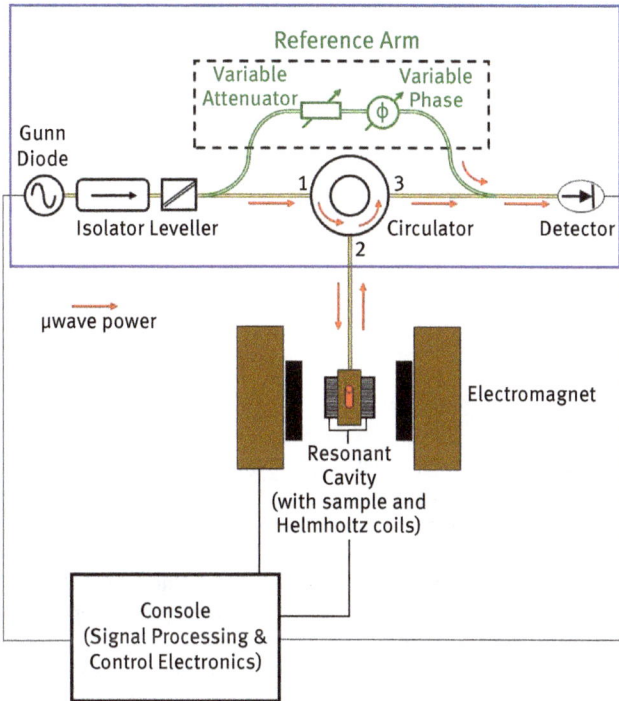

Fig. 6.1: Schematic diagram of the EPR spectrometer. The microwave radiation is produced by a Gunn diode (or klystron) with a small amount diverted through the bypass arm to the detector. The remaining radiation travels down a series of waveguides, an isolator (protects source from reflected microwave power) and a leveler (controls the microwave power) through a circulator to the sample in the resonant cavity (red arrows). When the instrument is properly tuned, no microwave power is reflected back from the cavity. When the magnetic field is swept through resonance, microwave power is absorbed by the sample in the cavity and a small signal is reflected back from the cavity that travels through the circulator and is mixed with the signal from the bypass arm. The combined signal then travels to the detector and the electronics console for processing.

with an AC magnetic field (such as the microwave magnetic field inside the cavity), its magnetic susceptibility is complex given by

$$\chi = \chi' - i\chi'' \tag{6.1}$$

The minimum magnetic susceptibility that can be detected is expressed by the imaginary component χ'' of the AC susceptibility [1]:

$$\chi'' = \frac{1}{Q\eta\pi} \left(\frac{kT\Delta v}{P_0} \right)^{1/2} \tag{6.2}$$

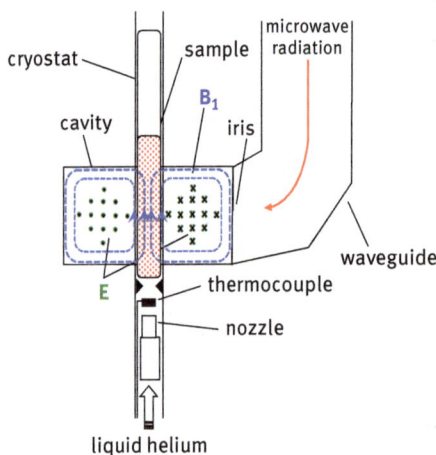

Fig. 6.2: Cross section of a rectangular TE_{102} cavity (see Section 6.2.7) with cryostat and sample. The microwave radiation is transmitted into the cavity through an adjustable circular iris and stored as standing waves. The magnetic field component of the microwave radiation ($\vec{B}_1$: dotted blue lines) is 90° out of phase relative to the electric field component ($\vec{E}$: circles and x's). The arrows show the direction of the $\vec{B}_1$ vectors inside the cavity. The $\vec{E}$ vectors are at right angles relative to $\vec{B}_1$, coming out of the page on the left half of the cavity (circles) and into the page on the right half of the page (x's). The external magnetic field, provided by an electromagnet, is perpendicular to the plane of the page. The shown configuration allows the detection of species with half-integer spins. In a bimodal cavity, the microwave frequency can be adjusted to generate a different configuration of the standing wave pattern inside the cavity that allows $\vec{B}_1$ to orient parallel to the external magnetic field for the detection of integer-spin systems.

where Q is the quality factor of the cavity (see Section 6.2.9), η is the cavity filling factor (see Section 6.2.9), T is the temperature of the sample, Δv is the bandwidth of the recording system and P_0 is the microwave power inside the cavity. High sensitivity requires Q to be high, η close to unity and the power as high as possible without causing saturation (see Section 3.6 for a discussion of saturation effects on the EPR signal). The bandwidth Δv is the easiest parameter to control with small values leading to high sensitivity. Small values of Δv mean that other instrumental parameters such as the microwave power, the magnetic field and the microwave frequency must not fluctuate at all during the recording of the signal. This implies that instrument stability is essential in EPR spectroscopy.

Commercial spectrometers employ microwave bridges and other components that are factory sealed allowing no access to the instrumentation inside. Anyone carrying out EPR spectrometers, however, should have a rudimentary knowledge of how the instrumentation works in order to better understand how spectra are generated and how to improve the data. In addition, many laboratories make modifications to commercially available spectrometers in order to carry out specialized measurements outside the limitations imposed by the commercial designs [2]. This section lists the main spectrometer components and briefly explains the function and operation of each component.

6.2.1 Microwave oscillators

As mentioned earlier, EPR spectrometers require a stabilized source of monochromatic microwave radiation. The majority of early EPR spectrometers accomplished this with a reflex klystron as the source of microwave radiation but the Gunn diode, which operates at lower voltages, has replaced the klystron in newer spectrometers [3].

The reflex klystron works by accelerating electrons to a high voltage then reflecting them back to the source and bunching them in such a way so that they emit radiation. Klystrons require frequency stabilization, which is achieved via the use of automatic frequency control, where the klystron frequency is compared to a standard frequency. When the klystron frequency drifts, an error signal is produced, amplified and applied to the klystron reflector voltage to bring the klystron frequency back to the correct value.

The Gunn diode is a solid-state oscillator, typically a gallium arsenide semiconductor, which produces microwave oscillations when biased above a certain voltage threshold [4]. It has a broad tuning range, and low noise can be achieved by phase locking to an external resonant circuit or cavity [5]. Another type of solid-state oscillator with low-noise characteristics found in some commercial spectrometers is one based on an yttrium–iron–garnet oscillator.

6.2.2 Propagation of microwave radiation in waveguides

Microwave radiation is delivered to the sample via a network of metal waveguides. These are typically rectangular in cross section and are constructed of brass with a layer of silver or gold on the inner surface to improve conductivity.

Electromagnetic radiation bounded by metallic walls behaves differently than radiation which freely propagates through space due to the multiple reflections that occur with the walls. The electromagnetic waves must also satisfy certain boundary conditions: for a waveguide made out of a perfectly conducting material, the electric and magnetic field components $\vec{E}$ and $\vec{B}$ must both be equal to zero. This requirement gives rise to the boundary condition that the $E_{\parallel}$ and $B_{\perp}$ must be zero on the inside wall. For monochromatic radiation propagating down the long dimension of the waveguide (along the z-axis), the electric and magnetic field components are given by

$$\vec{E}(x,y,z,t) = \vec{E}_0 e^{i(kz-\omega t)} \tag{6.3a}$$

$$\vec{B}(x,y,z,t) = \vec{B}_0 e^{i(kz-\omega t)} \tag{6.3b}$$

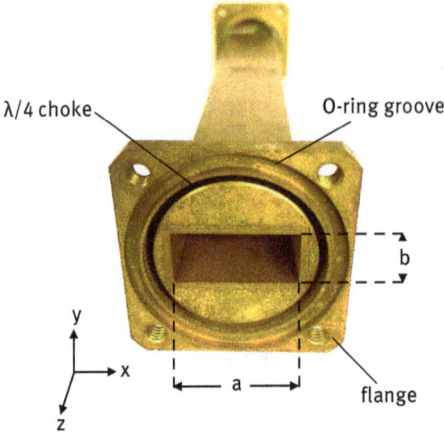

Fig. 6.3: X-band rectangular brass waveguide. This particular waveguide has a highly polished inner surface. The flange includes a groove for a rubber O-ring to allow for pumping on the system of waveguides and a λ/4 choke which prevents reflected radiation at the flange from traveling back into the waveguide. Cross-sectional dimensions: $a = 22.86$ mm and $b = 10.16$ mm. The direction of propagation is along the z-axis.

where k is the wave number ($k = 2\pi/\lambda$, λ = wavelength) and ω is the angular frequency ($\omega = 2\pi\nu$, ν = frequency) of the radiation. These fields must satisfy Maxwell's equations inside the waveguide:

$$\vec{\nabla} \cdot \vec{E} = 0 \rightarrow \vec{\nabla} \times \vec{E} = -\frac{\partial \vec{B}}{\partial t} \tag{6.4a}$$

$$\vec{\nabla} \cdot \vec{B} = 0 \rightarrow \vec{\nabla} \times \vec{B} = \frac{1}{c^2}\frac{\partial \vec{E}}{\partial t} \tag{6.4b}$$

The problem is reduced to finding expressions for $\vec{E}_0$ and $\vec{B}_0$. In general,

$$\vec{E}_0 = E_x\hat{x} + E_y\hat{y} + E_z\hat{z} \tag{6.5a}$$

$$\vec{B}_0 = B_x\hat{x} + B_y\hat{y} + B_z\hat{z} \tag{6.5b}$$

when $E_z = 0 \Rightarrow$ transverse electric (TE) waves
$\quad B_z = 0 \Rightarrow$ transverse magnetic (TM) waves
$\quad E_z = B_z = 0 \Rightarrow$ transverse electromagnetic waves (cannot propagate in hollow waveguides)

The predominant modes used in EPR spectroscopy are the TE_{mn} modes, TE_{10} in particular.[1] The magnetic field in this case that satisfies the above conditions is given by

$$B_z = B_0 \cos\left(\frac{m\pi x}{a}\right) \cos\left(\frac{n\pi y}{b}\right)$$

(6.6)

where a and b are the cross-sectional dimensions of the waveguide ($a < b$). The wave number k satisfies the expression

$$k = \left[\left(\frac{\omega}{c}\right)^2 - \pi^2\left[\left(\frac{m}{a}\right)^2 + \left(\frac{n}{b}\right)^2\right]\right]^{1/2}$$

(6.7)

The lowest frequency that can be propagated by the waveguide is called the cutoff frequency and is given by

$$\omega_{mn} = c\pi \left[\left(\frac{m}{a}\right)^2 + \left(\frac{n}{b}\right)^2\right]^{1/2}$$

(6.8)

The phase velocity in the waveguide is given by

$$v = \frac{\omega}{k} = \frac{c}{\left[1 - \left(\frac{\omega_{mn}}{\omega}\right)^2\right]^{1/2}}$$

(6.9)

which is larger than c but the wave energy propagates at the same speed as the group velocity

$$v_g = \frac{1}{dk/d\omega} = c\left[1 - \left(\frac{\omega_{mn}}{\omega}\right)^2\right]^{1/2}$$

(6.10)

which is less than c.

Electromagnetic waves traveling in waveguides will transmit energy along the length of the waveguide only when the frequency is above a certain value called the cutoff frequency. Below this value, the electric and magnetic field components of a wave sent down the waveguide will exponentially decrease and no energy will travel down the waveguide. The cutoff frequency depends on the mode, the type of waveguide and its dimensions. The corresponding wavelength is called the cutoff wavelength λ_c. For a rectangular waveguide of the type shown in Fig. 6.3, λ_c is given by [6]

$$\lambda_c = \frac{2}{\sqrt{\left(\frac{m}{a}\right)^2 + \left(\frac{n}{b}\right)^2}}$$

(6.11)

---

1 The subscripts mn indicate the number of half-wavelengths along the xy-coordinates as defined in Fig. 6.3. For the TE_{10} mode, there is one half-wavelength along the x-axis.

This equation holds for both the TE_{mn} and TM_{mn} modes.

When electromagnetic waves strike the surface of a conductor, they penetrate into the conductor and setup currents within a thin layer of the surface. The waves are attenuated inside the conductor with the amplitude exponentially decreasing relative to the amplitude in air. Considering waves propagating along the z-axis incident on a conductor perpendicular to the z-axis at $z = 0$, the amplitudes of the electric and magnetic field components inside the conductor are given by [7]

$$\vec{E}_0(z) = \vec{E}_0 e^{-\kappa z} \tag{6.12a}$$

$$\vec{B}_0(z) = \vec{B}_0 e^{-\kappa z} \tag{6.12b}$$

With

$$\kappa = \omega \sqrt{\frac{\varepsilon \mu}{2} \left(\sqrt{1 + \left(\frac{\sigma}{\varepsilon \omega} \right)^2} - 1 \right)^{1/2}} \tag{6.13}$$

where $\vec{E}_0$, $\vec{B}_0$ are the amplitudes in air, $\omega = 2\pi\nu$, and ε, μ, σ are the permittivity, permeability and conductivity of the conductor, respectively. The distance inside the conductor where the amplitude of the electromagnetic radiation falls to $|\vec{E}_0|/e$ and $\vec{B}_0/e$ is called the skin depth δ and is given by

$$\delta = \frac{1}{\kappa} \tag{6.14}$$

In order for waveguides to be able to support the propagation of the microwave radiation, they need to have a thickness that is larger than the skin depth.

6.2.3 Microwave isolator

A typical EPR spectrometer consists of a number of microwave waveguides and components that are connected to each other via coupling joints and flanges. These connections create the potential for spurious reflections of microwave radiation. The isolator is placed right after the microwave source in order to protect the source from such reflections of microwave power that might reach the source and possibly cause some damage or cause it to malfunction. The isolator is a nonreciprocal device that allows practically all the microwave radiation to go through in the forward direction and strongly attenuates any radiation that flows in the return direction. It consists of a ferrite core and employs the Faraday effect to prevent reflected radiation from reaching the microwave source. The electron spins in the ferrite core are aligned by an external magnetic field in such a way that causes a rotation of the plane of polarization of the microwave radiation relative to the incident polarization. In the most common setup, the linearly polarized incident radiation is rotated by 45° after going through the isolator. Any reflected radiation that comes back to the isolator is rotated by a further 45° that creates an overall 90° rotation relative to the incident polarization. The reflected radiation is consequently 90° out of phase with the plane of polarization of the waveguide and is completely attenuated and cannot reach the microwave source. This process is shown schematically in Fig. 6.4.

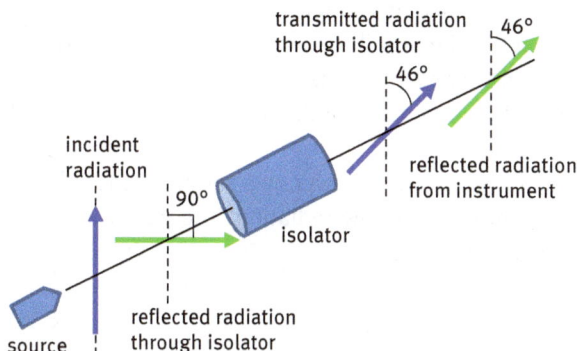

Fig. 6.4: Microwave radiation propagating from the microwave source toward the resonant cavity (blue) linearly polarized in the vertical direction. Plane of polarization is rotated by 45° after passing through the isolator. Radiation reflected back from the rest of the spectrometer (green) is rotated for an additional 45° after passing through the isolator.

6.2.4 Power leveler

The EPR phenomenon is only possible when a difference in the populations of the magnetic states of the spin system is maintained during the experiment. If large microwave powers are absorbed by the spin systems, then the populations of the two states are equalized, and no more absorption of microwaves can occur. In such a case, the system is saturated and an EPR signal can no longer be observed. The power leveler is inserted in the microwave circuit to prevent such saturation of the sample. This device is a power attenuator that cuts the microwave power emitted by the source down to a constant value, typically 200 mW.

6.2.5 Reference arm: attenuators and phase control

The purpose of the reference arm is to divert some of the power from the source and send it to the diode detector in order to bias it into the linear region for proper signal detection. The power is diverted from the main waveguide via a directional coupler. The directional coupler employs a side arm that takes a small amount of the forward microwave power and diverts it into another waveguide that forms the reference arm. The amount of diverted power P_d relative to the incident power P_i in decibels is given by

$$C(\mathrm{dB}) = 10\log_{10}\frac{P_i}{P_d} \tag{6.15}$$

A typical value of directional couplers employed in a reference arm is $C \approx 20$ dB. The signal that travels through the reference arm enters the main waveguide past the circulator and gets mixed in with the signal reflected from the cavity. In order to properly mix the two signals, the amplitudes and phases must be set so that there are no out–of-phase components. This is achieved via the use of an attenuator and a phase shifter.

An attenuator reduces the incident microwave power by a certain amount caus-ing a concomitant reduction in the amplitude of the radiation. The power attenua-tion in decibels produced by this element is given by

$$\mathcal{A}(\mathrm{dB}) = 10\log_{10}\frac{P_i}{P_t} \tag{6.16}$$

where P_i and P_t are, respectively, the incident and transmitted powers through the attenuator. Attenuators can be inserted in series anywhere in the microwave circuit. They come as either a calibrated attenuator that reduces the power by a predeter-mined amount or with a dial that can be set to the desired attenuation. The de-crease in power is achieved through the insertion of a small dielectric element into the broad side of the waveguide. The dielectric element absorbs microwaves and depending on how far it extends into the waveguide determines the amount of ab-sorbed power.

A phase shifter works by altering the long dimension (a) of the cross section of the waveguide. This is possible due to the fact that the guide wavelength depends on this dimension according to

$$\lambda_g = \frac{\lambda}{\left[1 - \left(\dfrac{\lambda}{2a}\right)^2\right]^{1/2}} \tag{6.17}$$

where λ is the free wavelength in air. Changing a causes a change in λ_g which affects the wave patterns inside the waveguide and results in a phase shift of the waves.

6.2.6 Circulator

The circulator is a nonreciprocal device with several ports that allow microwave ra-diation to enter and leave the ports in a certain manner. Early circulators typically had four ports but three ports are employed in the most practical setup used in current spectrometers. The purpose of the circulator is to direct microwave power from one port to the next in a sequential manner: microwave power that enters port 1 is directed to port 2 and travels into the tuned cavity. Power reflected from the cavity during resonance comes back to port 2 and is directed to port 3 and onto the detector. The length the microwaves travel from port 1 to 2 and from 2 to 3 is equal to $\lambda_g/4$, making sure that any microwaves that get backward reflected from either 3 to 2 or 2 to 1 destructively interfere with the incident microwaves to preserve the nonreciprocal nature of the device. The scattering matrix obeyed by the three-port circulator is given by

$$\begin{bmatrix} B_1 \\ B_2 \\ B_3 \end{bmatrix} = \begin{bmatrix} 0 & 0 & 1 \\ 1 & 0 & 0 \\ 0 & 1 & 0 \end{bmatrix} \begin{bmatrix} A_1 \\ A_2 \\ A_3 \end{bmatrix} \tag{6.18}$$

where A_i are the input and B_i the output microwaves at port i [3].

6.2.7 The resonant cavity

The resonant cavity allows large microwave energy densities to form within its volume. This is where the sample is placed in a position of maximum microwave magnetic field and minimum microwave electric field. The microwave energy is stored in the form of standing waves. The standing waves allowed by the boundary conditions are TE (TE_{mnp} modes) and TM (TM_{mnp} modes). As in the case of the traveling waves in the waveguide, the subscripts mnp indicate the number of half-wavelengths along each of the coordinate axes. In cavities, since we have standing waves we use an extra subscript to indicate the number of half-waves along the z-axis as well (see Fig. 6.5 for the orientation of the coordinate axes).

Cavities are made out of the same material as the waveguide and are typically gold or silver plated to achieve high energy densities. Two types of resonant cavities are common in EPR spectrometers: rectangular and cylindrical. Rectangular cavities are more commonly used at X-band, whereas cylindrical cavities are more common at Q-band frequencies for reasons that will become apparent below.

6.2.7.1 Rectangular cavity
This cavity is essentially a section of a rectangular waveguide that is terminated by two flat metallic walls perpendicular to the long dimension of the waveguide. Its function is to concentrate the microwave radiation at the sample location. This is achieved by letting the microwave radiation into the cavity through an aperture (the coupling iris) where it sets up standing waves inside the cavity. The geometry of the cavity is shown in Fig. 6.5, where the terminating end walls are perpendicular

to the long dimension d. The resonant frequency ν_0 of a rectangular cavity operating in the TE_{mnp} mode is given by [3]

$$\nu_0 = \frac{1}{2\pi\sqrt{\mu\varepsilon}} \left[\left(\frac{m}{a}\right)^2 + \left(\frac{n}{b}\right)^2 + \left(\frac{p}{d}\right)^2 \right]^{1/2} \tag{6.19}$$

where μ and ε are the permeability and permittivity, respectively, of the material in the cavity (typically air). The cutoff wavelength (maximum wavelength that can be supported) for such a cavity is given by eq. (6.11).

(a)

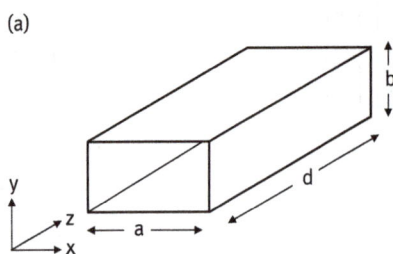

(b)

Fig. 6.5: (a) Geometry of a rectangular cavity. Typical inside dimensions for full-wavelength TE_{102}-mode X-band cavities are $a = 22.9$ mm, $b = 10.2$ mm and $d = 2a$. (b) Rudimentary homebuilt TE_{102} rectangular cavity attached to the same type flange as the one shown in Fig. 6.3. In most commercial cavities, the modulation coils (Section 6.2.9) are symmetrically attached to the broad face A and its corresponding parallel face. In addition, parallel slots are sometimes cut into the B face to allow irradiation of the sample from an external source. The inside surface of the cavity is usually plated with a thin layer of gold for improved conductivity but primarily to prevent any metal impurities from the brass interfering with the EPR signal.

The electric and magnetic field components of the microwave radiation inside the cavity are not in-phase as in traveling electromagnetic waves but they are 90° out of phase. This is advantageous for EPR experiments since the sample must be placed at a location where the microwave magnetic field is large but the microwave electric field is zero to prevent dielectric losses due to nonresonant absorption of microwave power and heating of the sample. The TE_{102} cavity mode is the preferred cavity mode used with rectangular X-band resonators. The distribution of the microwave magnetic and electric field components is shown in Fig. 6.6.

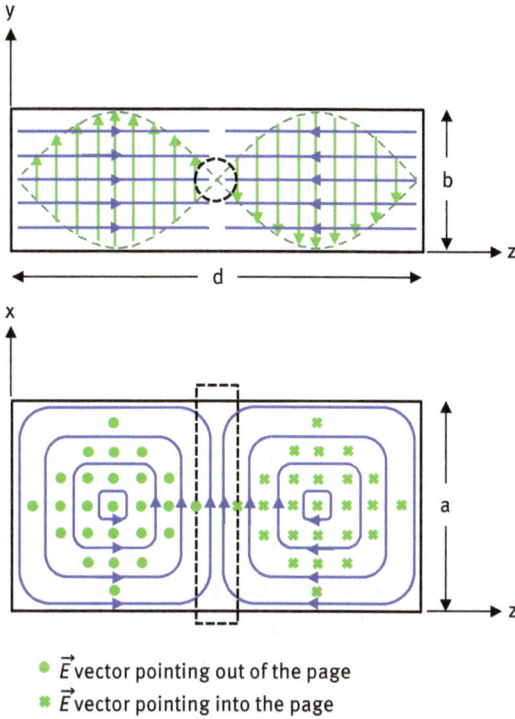

- ● $\vec{E}$ vector pointing out of the page
- ✱ $\vec{E}$ vector pointing into the page

Fig. 6.6: Two perspectives of the distribution of the electric (green) and magnetic field (blue) components of the microwave electromagnetic field inside a rectangular cavity in the TE_{102} mode. The dashed lines indicate the position of a sample EPR tube which is placed in a position of maximum $\vec{B}_1$ and minimum $\vec{E}$.

The solutions of the wave equation subject to the boundary conditions that give the distribution of the microwave electromagnetic field shown in Fig. 6.6 are given by

$$B_x = \frac{B_0}{\left[1 + \left(\dfrac{d}{2a}\right)^2\right]^{1/2}} \sin\frac{\pi x}{a}\cos\frac{2\pi z}{d} \qquad (6.20a)$$

$$B_z = \frac{-B_0}{\left[1 + \left(\dfrac{d}{2a}\right)^2\right]^{1/2}} \cos\frac{\pi x}{a}\sin\frac{2\pi z}{d} \qquad (6.20b)$$

$$E_y = i\left(\frac{\mu}{\epsilon}\right)^{1/2} B_0 \sin\frac{\pi x}{a}\sin\frac{2\pi z}{d} \qquad (6.20c)$$

6.2.7.2 Cylindrical cavity

Cylindrical cavities have a higher Q than rectangular cavities with the same dimensions. They find applications in frequencies higher than K-band, in particular at Q-band, where the size limitations due to the smaller microwave wavelengths make rectangular cavities difficult to work with. They can support both TE and TM modes, with the TE_{011} mode the most prevalent mode in use. The equations that describe the distribution of the magnetic and electric fields inside the cavity are given in terms of Bessel functions, and the magnetic field distribution is shown schematically in Fig. 6.7 for two different coupling configurations.

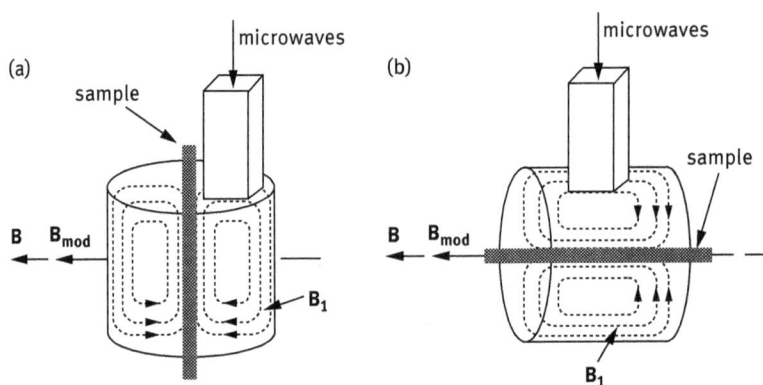

Fig. 6.7: Comparison of coupling methods to the TE_{011} mode in a cylindrical cavity: (a) conventional end-wall coupling, (b) cylindrical wall coupling. **B** = applied magnetic field, B_{mod} = modulation field, B_1 = magnetic field component of microwave radiation. In both cases, **B** rotates in a plane at right angles to the page. [2]

A special type of cavity was developed by the author that operates at Q-band frequencies and can be used to study metalloproteins with weak EPR signals. It is especially useful for the study of systems containing non-Kramers ions with large zero-field splittings that in the past were classified as "EPR silent." This cavity employs a helical metal body that can sustain a TE_{011} mode and can also allow large modulation fields to penetrate into the sample to improve the signal-to-noise ratio. A side-wall coupling, as shown in Fig. 6.8, makes this cavity suitable for the detection of Kramers ions as well without having to change the resonant mode by simply rotating the external magnetic field.

6.2.7.3 Dual-mode cavities

With dual-mode cavities, two different modes can be excited at different resonance frequencies. Shifting the frequency switches between the two modes without the need to disturb the sample. There are two cavity dips that correspond to each mode. The usefulness of such cavities lies in the fact that the distribution of the microwave

Brass tapered
sections

Teflon Wedge

Coupling Iris   Epoxy Shell

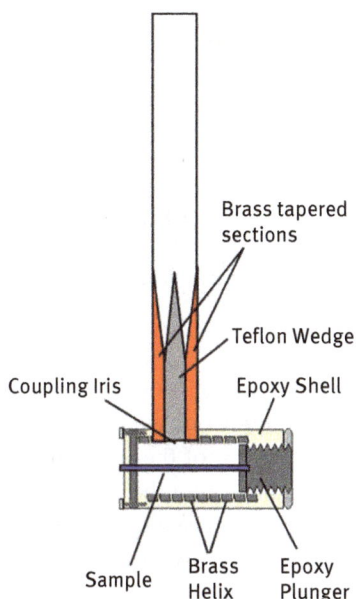

Fig. 6.8: Q-band helical cavity with side-wall coupling that employs a Gordon coupler. The brass helix is embedded inside an epoxy matrix and the inside of the cavity is gold plated for improved Q. A special epoxy is used that can withstand repeated cycling between liquid helium and room temperatures [2].

Sample   Brass   Epoxy
          Helix   Plunger

magnetic field is parallel to the external magnetic field in one mode and perpendicular in the other, allowing the study of both Kramers and non-Kramers systems without changing the location of the sample in the cavity. Commercial instruments are in use that employ cavities, such as the Bruker ER4116DM resonator, operating in both the TE_{102} ("perpendicular") and TE_{012} ("parallel") modes. Kramers ions can be detected when the cavity is tuned in the perpendicular mode and non-Kramers ions in the parallel mode.

6.2.8 Coupling the microwave radiation into the cavity

The resonant cavity concentrates the microwave power at the location of the sample. The amount of power that enters the cavity is controlled by the iris, an opening on one of the side walls of the cavity where a nylon screw with a metal tip is placed in front of it as shown in Fig. 6.9. This iris is circular and centered on the ab cross section of the waveguide. Different types of iris geometries such as rectangular and elliptical also exist depending on the nature of the coupling desired but the centered circular type is the most common type.

Depending on the placement of the metal tip in front of the aperture, microwaves are allowed to enter the cavity or are reflected back into the waveguide and/or cavity so that standing waves are set up inside the cavity. The iris allows us to "match" the cavity to the rest of the waveguide system for maximum power transfer. This follows from an electric circuit analogy of the microwave system where maximum electric

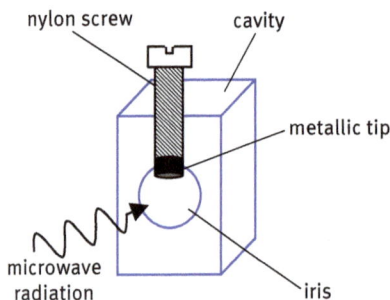

Fig. 6.9: Configuration of a centered circular iris that allows coupling of microwave radiation into a rectangular resonance cavity. Clockwise motion of the nylon screw moves the metal tip further in front of the iris.

power is transferred from the power source to a reactive load. In order to transfer maximum electric power from the power source to the RLC part of the circuit, the total impedance of the source must be equal to the impedance of the RLC circuit. When this happens, all the electric power from the source flows into the RLC circuit without any of it getting reflected back to the source. The microwave cavity can be thought of as an RLC circuit where the stored microwave magnetic field is represented by the inductive component, the stored microwave electric field by the capacitive component and the dissipative (ohmic) losses in the cavity walls by the resistive component. The iris in the shape of a circular aperture can be thought of as an inductive or capacitive element in series with the cavity RLC circuit. Proper adjustment of the nylon screw in front of the iris is equivalent to adjusting the impedance of the equivalent circuit element of the iris, allowing us to match the impedance of the cavity to the impedance of the rest of the microwave system. Maximum microwave power can then be transferred from the klystron or Gunn diode to the cavity without any power being reflected back from the cavity. This way, the only time that any microwave power gets reflected back from the cavity occurs when power gets absorbed by the sample when it goes through resonance. At this point, an unbalanced signal is produced that travels back from the cavity to the detector.

Another way to couple the microwave radiation into the cavity is achieved via the use of a Gordon coupler, which is used in many special-purpose instruments [2]. The Gordon coupler employs a tapered waveguide that brings the dimensions of the waveguide below cutoff. A Teflon wedge is inserted into the tapered part of the waveguide and brings it above cutoff allowing the microwaves to propagate through the Teflon wedge and into the iris as shown in Fig. 6.10. Raising the Teflon wedge creates an evanescent wave between the wedge and the iris. Adjusting the height of the wedge above the iris controls the amplitude of the evanescent wave that enters the cavity, thus controlling the microwave radiation entering the cavity.

Traveling waves in waveguides have magnetic and electric field components that are in-phase. When standing waves are formed, the magnetic and electric field components are 90° out-of-phase to satisfy the boundary conditions imposed by the interaction of the electromagnetic radiation with the metallic walls of the cavity.

microwave
radiation

waveguide

teflon wedge

tapered
waveguide

iris

microwave
magnetic field

cavity

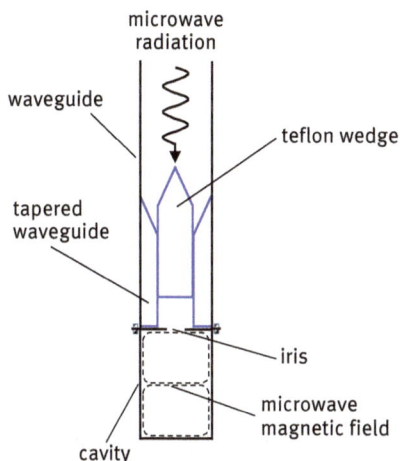

Fig. 6.10: Gordon coupler for a full-wavelength cavity. The iris is a circular aperture with a radius of a few mm. The Teflon wedge can be adjusted via a coupling rod external to the waveguide so that it does not interfere with the microwaves. Such an arrangement of cavity and coupler is suitable for single-crystal studies where various orientations of the crystal axes can be achieved relative to both $\vec{B}_1$ and the external magnetic field $\vec{B}$ [8].

In other words, $\vec{B}_1$ is maximum where $\vec{E}$ is zero and vice versa. The sample is placed at a position of maximum $\vec{B}_1$ as shown in Fig. 6.6. Care must be taken to use sample tubes that are small enough so that the sample does not extend into regions of finite $\vec{E}$. If that happens, nonresonant absorption of microwave power will occur that leads to heating of the sample and also to a decrease of the Q-factor of the cavity. Typical sample tubes used for powder or frozen solution samples at X-band frequencies have an inner diameter of 3 mm.

The cavity mode produced by a resonant cavity is shown in Fig. 6.11, where the reflected microwave power is displayed against the microwave frequency.

With the dip centered as shown in Fig. 6.11(c), the cavity can be critically coupled by adjusting the iris. With the cavity at critical coupling, as shown in Fig. 6.12, no power is reflected back from the cavity. As shown the cavity is undercoupled, where some of the microwave power is reflected back from the cavity. Turning the coupling screw counterclockwise moves the metal tip away from the iris causing ΔP to increase further undercoupling the cavity. Turning the coupling screw clockwise moves the metal so that it covers more of the iris causing ΔP to decrease and bringing the coupling closer to critical coupling, which occurs when ΔP is zero. If the iris is adjusted beyond critical coupling, ΔP starts to increase again bringing the resonator into the overcoupling regime.

6.2.9 The quality factor Q

The quality factor Q of the cavity is defined as the ratio of the energy stored in the cavity over the energy dissipated per cycle [3]:

$$Q = 2\pi \frac{\text{energy stored}}{\text{energy} \frac{\text{dissipated}}{\text{cycle}}} \tag{6.21}$$

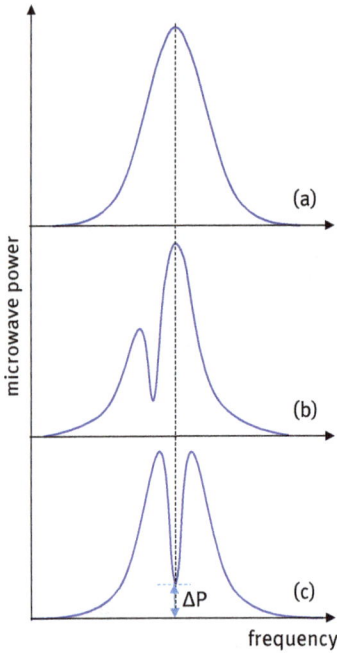

Fig. 6.11: (a) Typical cavity mode when instrument frequency is away from the cavity resonant frequency. (b) As the instrument frequency is adjusted, the cavity dip appears. (c) The instrument is tuned to the resonant frequency when the cavity dip is centered on the cavity mode. When the bottom of the dip is above the baseline, the instrument is either undercoupled or overcoupled.

The unloaded Q of a cavity depends on the particular mode and the cavity dimensions. It can be experimentally measured from the cavity mode at critical coupling as shown in Fig. 5.10. If the width of the cavity dip at half maximum is Δv, then Q can be determined from

$$Q = \frac{v_0}{\Delta v} \tag{6.22}$$

where v_0 is the resonant frequency. Typical Q factors range from 500 to 10^4. The Q-factor is related to the magnetic susceptibility according to eq. (6.2).

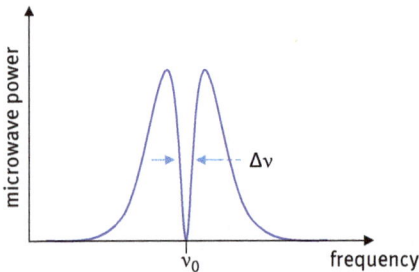

Fig. 6.12: Cavity mode at critical coupling. When critically coupled, there is no microwave power reflected back from the cavity.

Gold plating the interior of the cavity improves the Q. Plating with high-purity gold also prevents detection of impurities that might be in the brass walls of the cavity. As discussed in Section 6.2.2, electromagnetic radiation that penetrates into metallic conductors exponentially decreases, giving rise to a characteristic penetration depth that depends on the frequency of the radiation and the material of the metal conductor. The distance from the surface of the conductor to where the amplitude of the radiation decreases to $1/e$ its value outside the conductor is called the skin depth δ. Cavities must be thicker than the penetration depth of the microwaves so that they can support the storage of the microwaves and they must also be thinner than the penetration depth of the modulation frequencies so that the modulation field can reach the sample. When brass waveguides and cavities are gold plated for better conductivity, the gold layer must be thick enough to support the microwave mode but thin enough to allow penetration of the modulation field. For cavities, a gold layer of thickness in the μm range and a brass wall thickness in the mm range satisfy these conditions. The difference in thickness is due to the fact that δ is inversely proportional to the square root of the frequency (eq. (6.14)). For microwave frequencies (GHz range), δ is going to be much smaller than δ for modulation frequencies (kHz range) allowing the construction of cavities with thicker walls that provide structural stability.

The strength of the EPR signal is proportional to the product of Q and the filling factor η [3]. The filling factor represents the faction of the microwave energy in the cavity that is absorbed by the sample:

$$\eta = \frac{\int_{\text{sample}} B_1^2 dV}{\int_{\text{cavity}} B_1^2 dV} \tag{6.23}$$

The integrals can be evaluated for cavities with typical geometries (rectangular or cylindrical) and microwave modes for different types of sample sizes and geometries [3]. For rectangular cavities and samples in completely full tubes, where the sample dimensions are much smaller than the cavity dimensions making $\vec{B}_1$ constant over the entire sample, the filling factor is equal to the ratio of sample volume over the cavity volume:

$$\eta = \frac{V_s}{V_c} \tag{6.24}$$

Consequently, a larger volume sample will have a larger η but this does not lead to a larger signal, since a larger sample leads to a lower Q and also violates the requirement that $\vec{B}_1$ be constant over the sample volume. Therefore, the Q and η must be optimized for maximizing the signal.

6.2.10 Modulation coils

The modulation coils provide the oscillating magnetic field that is necessary for detecting the EPR signal. This magnetic field is superimposed onto the static magnetic field and sweeps through the EPR absorption line as the static magnetic field is swept through resonance. The coils that provide this magnetic field are built in a Helmholtz arrangement to provide a uniform magnetic field in the region between the coils. In this arrangement, the coils share a common axis of symmetry with the radius of the coils equal to the distance between the coils $r = a$ as shown in Fig. 6.13. The magnetic field produced is very uniform along the axis of symmetry of the coils in the region between the coils with dB/dx and d^2B/dx^2 both equal to zero at the midpoint of the coils where the sample is placed. Using the geometry of Fig. 6.13, the magnetic field along the x-axis in Tesla for a Helmholtz coil with N turns of wire carrying a current I is given by

$$B = \frac{\mu_0 N I a^2}{2} \left(\frac{1}{\left[\left(x + \frac{a}{2} \right)^2 + a^2 \right]^{3/2}} + \frac{1}{\left[\left(x - \frac{a}{2} \right)^2 + a^2 \right]^{3/2}} \right) \tag{6.25}$$

Sometimes design parameters do not allow an exact Helmholtz arrangement to be achieved but as long as $r \approx a$, then the uniformity of the modulation field between the coils is maintained. The coil winding is made out of Litz wire[2] with enough turns to provide fields in the tens of Gauss. In most commercially available spectrometers, the coils are mounted directly on the walls of the cavity. For rectangular cavities, the coils are placed in recesses against two parallel cavity sides and in cylindrical cavities they are mounted so that they curve around the lateral surface of the cavity. Modulation coils can also be mounted away from the cavity as in the case of immersion cryostats in order to avoid boil off of the cryogens.

The modulation coils are typically driven at audio frequencies with 100 kHz providing a balance between enough penetration of the modulation field through the walls of the cavity and low signal noise (because of the $1/f$ crystal noise, higher frequencies improve the detector noise).

Effect of modulation on the EPR signal: As the external magnetic field is slowly swept through resonance, the modulation field is swept over the shape of the EPR absorption line as shown in Fig. 6.14. The modulation amplitude must be a fraction of the linewidth of the EPR absorption line; otherwise, distortion and broadening of the signal will occur. Sweeping the modulation field in this way produces a response

---

2 Litz wire is a special type of wire that is made of multiple individually insulated strands of wire that are twisted or braided together. Because of this design, they reduce AC losses making them suitable for AC circuits operating at the low end of the radio frequency range.

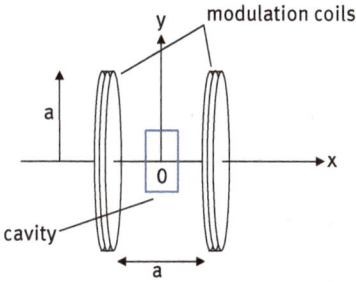

Fig. 6.13: Modulation coils in a Helmholtz arrangement. For coils of N turns with a current I flowing through in the geometry shown here, the magnetic field along the x-axis in Tesla is given by eq. (6.25). The cavity is centered at the origin of coordinates.

that is sensitive to the slope of the absorption line at the various field values: the response will increase as the field is swept through the rising part of the absorption line, rising to a maximum at the inflection point and then decreasing to zero at the peak of the absorption line. It then decreases to negative values as the slope becomes negative on the decreasing side of the absorption line until it reaches a minimum value at the inflexion point, and then the response increases until it reaches zero. This change in phase as modulation is swept through the maximum of the absorption line is indicated by the two modulation inputs shown in Fig. 6.14. The resulting signal is fed to the crystal detector and then through a narrow band amplifier to a phase-sensitive detector with a reference signal fed directly from the modulation source. The result is a derivative signal with good signal-to-noise ratio due to the rejection of noise by the narrow band amplifier.

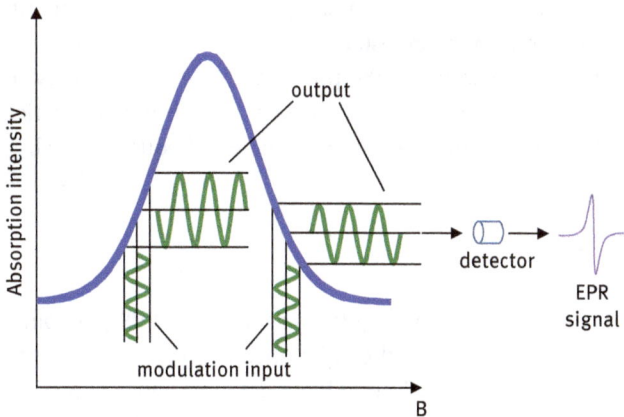

Fig. 6.14: Field modulation, shown in the vertical direction, is swept through the absorption line (blue trace) producing the horizontal EPR signal. Notice the phase change in the signal on either side of the maximum. Signal amplitude changes as the slope of the absorption line changes.

It is obvious from the previous discussion that a larger modulation amplitude will produce a larger signal and increase the sensitivity of the instrument. It does also lead, however, to a distortion and broadening of the signal as mentioned above since the modulation cannot now follow the true derivative of the absorption line. Excessive broadening of the EPR line also leads to loss of resolution, especially in spectra with many closely spaced EPR lines. A rule of thumb is to typically keep the modulation amplitude at ~ 1/10 of the linewidth. Sometimes, however, it is necessary to ignore this rule of thumb and send as large a modulation amplitude as possible into the cavity. This is true for many metalloprotein samples which typically have very weak and broad EPR lines. Increasing the modulation amplitude to large values allows better detection of such lines and greatly improves the sensitivity of the instrument. Such methods are employed in homebuilt instrumentation (such as the modulation coils used in the Q-band spectrometer in the Hendrich Lab, Department of Chemistry, Carnegie Mellon University) where large modulation coils are installed outside the immersion cryostat containing the cavity [2]. These coils are specially designed to allow large powers to be applied from the modulation amplifier (up to ~ 500 W) by incorporating water-cooling methods to prevent the coils from overheating and shorting out. Very weak signals have been detected using this instrument such as the ones from Photosystem II [9]. These coils are shown in Fig. 6.15. Due to these coils being located a bit far from the cavity, they are in an approximate Helmholtz arrangement that produces a field that is quite constant at the sample. In order to provide enough penetration of the modulation field into the cavity, the tail of the cryostat had to be specially designed with slits cut into the metallic part of the tail, and the cavity was also made out of a helical metallic body embedded inside a cylindrical body made out of epoxy. For this reason too, these modulation coils also operate at lower modulation frequencies (usually 12.5 kHz instead of the 100 kHz which is typical in commercial instruments) that allow better penetration into the cryostat and cavity. Careful matching of the impedance of the coils to the impedance of the power amplifier is necessary to avoid reflections of power into the amplifier. This is achieved through parallel–series circuits of capacitors that each operate at the particular modulation frequency used.

6.2.11 Electromagnet

The magnetic field that gives rise to Zeeman splitting that produces the EPR phenomenon is provided by an electromagnet whose field can be linearly swept from zero to typically about 1 T. The magnetic fields required for EPR are not large, since at X-band frequencies, $g = 2.0$ signal appears at a magnetic field $B_0 \approx 0.33$ T (3.3 kG) and at Q-band frequencies at $B_0 \approx 1.25$ T (12.5 kG). It does, however, require highly stable and homogeneous magnetic fields which make the electromagnets quite costly. The magnetic field must be homogeneous over the volume of a sample; otherwise, broadening of the EPR line will result since parts of the sample could be at slightly different magnetic fields. Stability is important to ensure that the magnetic field does not drift during signal acquisition which also leads to artificial broadening of the EPR line. Typical

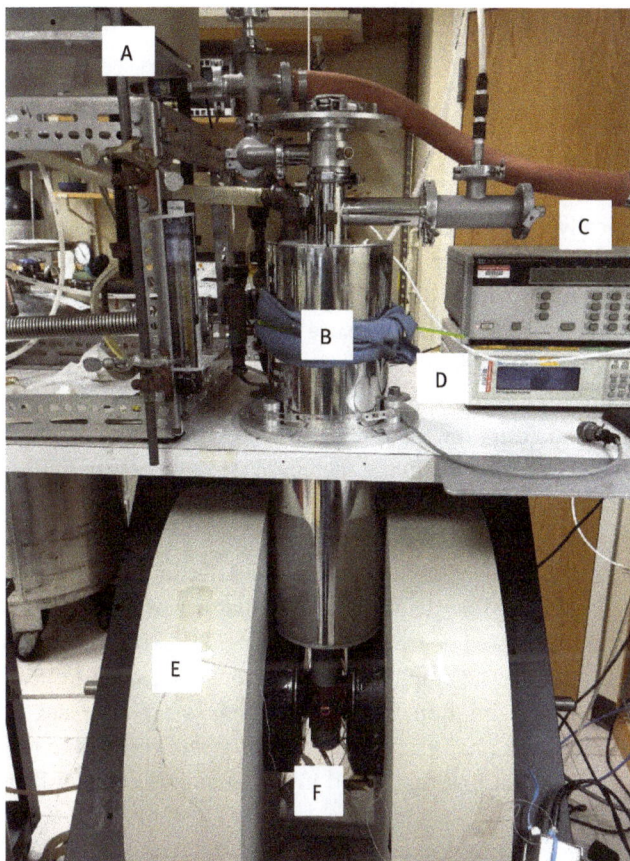

Fig. 6.15: Q-band spectrometer with a specially designed probe inside an immersion cryostat.
A, Bruker bridge; B, immersion cryostat (CRYO Industries); C, frequency counter; D, temperature
controller; E, electromagnet with tapers; F, modulation coils on frame wrapped around the cryostat tail.
(Hendrich lab, Department of Chemistry, Carnegie Mellon University)

relative homogeneities $\Delta B_0/B_0$ required for EPR are of the order of one part per million
(ΔB_0 is the magnetic field variation over the sample volume) [3]. A mapping of the magnetic field produced by the electromagnet within the magnet gap can be done to verify
the homogeneity of the field by using an NMR probe to take measurements to cover the
volume that the microwave cavity occupies.

Typical electromagnets have tapered cylindrical pole caps with a diameter of 9–12″
and a separation of ~ 3″. The field is regulated by a Hall probe that is attached to one of
the pole caps and connected to the electronics console. Proton NMR magnetometers
can be used to calibrate the magnetic field [3]. Corrections to the magnetic field can
also be made by recording the signal from a sample of the free radical diphenyl-

picrylhydrazyl, which contains a highly delocalized electron in each molecule producing a very strong sharp signal with $g = 2.0023$.

6.2.12 Detector

Microwave detectors are typically diodes that act as microwave rectifiers: they convert microwave power into an electric current that can be amplified and detected. The typical output of these diodes is 1 mA/mW. Diodes are nonlinear devices that work best when operating in the linear region (Fig. 6.16(a)). The use of a bypass arm as a component of a microwave bridge offers the advantage of adjusting the microwave power that falls on the detector independently of the amount of power that interacts with the sample inside the cavity. Consequently, the microwave power reaching the detector can be adjusted to bias the detector into the linear region. Microwave powers in the mW region are required for this purpose.

A lock-in amplifier is a phase-sensitive detector that detects signals of a certain frequency and phase by comparing the signal from the crystal detector to a reference signal. The reference signal is supplied by the same source that drives the modulation coils, locking the EPR signal to the same frequency allowing the EPR signal to be detected and thus suppressing noise. The amplifier has a narrow bandwidth for this purpose that is centered on the modulation frequency, which is typically set at 100 kHz. Since a lock-in amplifier is a phase-sensitive detector, it will respond to the different slopes of the EPR absorption line that is fed into it giving the typical first derivative EPR line as its output.

6.3 Cryogenic systems

Carrying out EPR experiments at low temperatures is desirable for many reasons. The most important reason is for systems where there is a strong spin–lattice relaxation coupling. Signals from these systems cannot be observed at room temperatures since the broadening of the EPR lines is so large that makes them indistinguishable from the signal noise. Systems with rare earth ions fall within this category since such ions exhibit strong spin–orbit coupling which gives rise to fast spin–lattice relaxation times. Such systems need to be studied at very low temperatures, close to or at liquid helium temperatures. In certain cases, the EPR lines from rare earth systems get broad enough even at 20 K to make them difficult to detect. Samples dissolved in many liquid solvents, especially water, need to be frozen before they can be studied. This is due to the strong absorption of microwave power by liquid solvents with large dielectric constants. The best example is water, which has a large electric dipole moment in liquid form which becomes much smaller when frozen resulting in smaller dielectric losses in the cavity. This allows the use of larger diameter EPR tubes

(a)

(b)

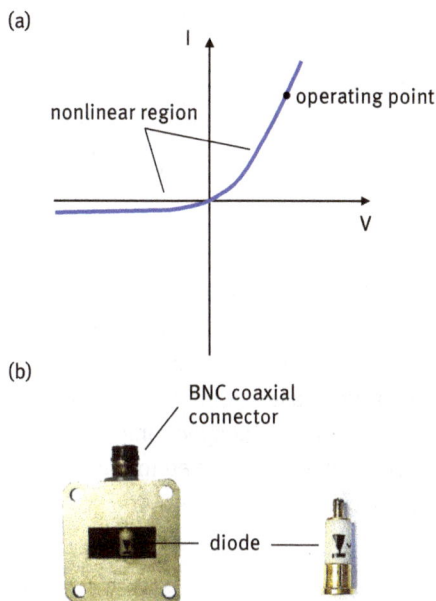

Fig. 6.16: (a) Current–voltage graph for a diode. A nonlinear relationship exists at moderate positive voltages. At higher voltages, the relationship is linear and the diode has a more ohmic behavior. The detector will properly operate when the diode is biased onto this region. (b) K-band waveguide with a coaxial connector that couples to the detector which can be seen in protruding into the waveguide. The waveguide is terminated with a flat wall behind the diode.

(typically 3 mm inner diameter for X-band experiments) increasing the volume of the sample in the cavity resulting in an increase in signal intensity, which is desirable for many biological systems that have notoriously weak EPR signals [10]. Air-sensitive samples also benefit from low temperatures since they need to be carefully handled to avoid oxidation of the sample. Once these samples are frozen inside EPR tubes, they cannot easily be oxidized when exposed to air and can be readily transferred from a storage Dewar to the EPR cryostat (see Section 6.4).

Early cryostats were immersion-type glass Dewar systems capable of reaching liquid helium temperatures. These have evolved into stainless steel Dewars but are not widely used in EPR due to the elaborate gas and vacuum systems that are required to bring the sample into thermal contact with the cryogen. The complexity of the setup does not allow for quick replacement of samples in the cryostat. Continuous-flow cryostats were developed that utilize cold helium or nitrogen gas to cool the sample. These cryostats allow quick sample replacement, enabling researchers to scan dozens of samples in a matter of hours. These have become the standard cryostats employed in most EPR spectrometers.

6.3.1 Continuous-flow systems

Continuous-flow cryostats employ differences in pressure to pull liquid helium out of an external storage Dewar, through a flexible transfer line, and pass it around the sample which is situated inside a cold finger. The Oxford Instruments ESR 900 cryostat shown in Fig. 6.17 works over the temperature range 4.2–300 K, while cryostats that employ a liquid helium reservoir can go down to temperatures of around 1.9 K by reducing the vapor pressure of the liquid helium bath (such as the Oxford Instruments ESR 910 cryostat). The flow rate of the cold helium gas is controlled by a needle valve that sends the helium through a nozzle below the sample tube and allows for precise temperature control. The consumption of liquid helium at very low temperatures ($T < 10$ K) varies between 1 and 2 L/h but decreases significantly for higher temperatures. A cryostat of this type is shown as part of an EPR spectrometer assembly in Fig. 6.18. Also shown is a liquid helium dewar which is attached to the cryostat via a flexible transfer line during low-temperature experiment.

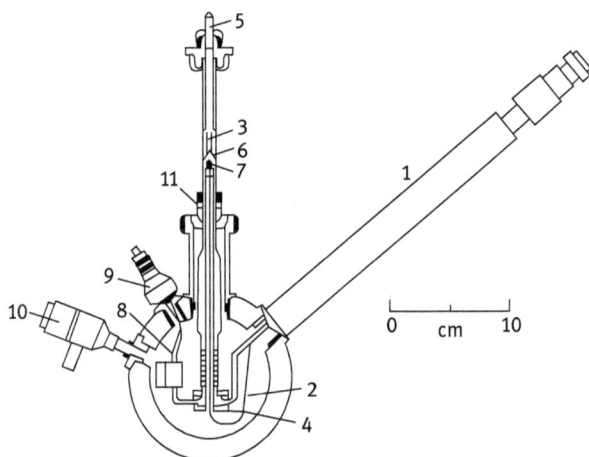

Fig. 6.17: Schematic of a continuous-flow cryostat for EPR spectrometers developed by the Oxford Instrument Company. 1, transfer tube arm; 2, capillary tube; 3, sample region; 4, central junction point; 5, O-ring seal for sample loading; 6, thermocouple; 7, heater; 8, 9, vacuum feedthroughs; 10, pressure relief valve; 11, threaded bushing [11].

A thermocouple right above the nozzle provides direct measurement of the temperature at that location. A heater in the form of a doubly-wound coil is wrapped right below the nozzle. The heater and the temperature sensors are monitored by a temperature controller that maintains the temperature of the system to set values. Because the nozzle is situated a few centimeters below the sample tube, there can be a temperature gradient created between the nozzle and the sample, especially at higher temperatures where the helium flow is slow. To avoid such discrepancies, it is customary to insert a small carbon-glass thermometer inside an EPR tube and use it in situ to calibrate the temperature at the sample location, typically at the beginning of a series of measurements. Another method of determining the sample temperature with a relatively high degree of accuracy is being pursued at the author's laboratory, which is particularly useful for frozen solution metalloprotein samples. This method uses single-crystal lanthanum ethylsulfate doped with very small amounts of Ce(III) ions. The $J = 5/2$ spin manifold of Ce(III) in this complex is split into three Kramers doublets with very well-known energy splittings. Very strong sharp signals with small linewidths are obtained at low temperatures with intensities that depend on the sample temperature. Measurement of the relative intensities of the EPR signals can yield accurate measurements of the temperature of the crystal. When these crystals are embedded inside a frozen solution sample, they can provide the temperature of the sample. Because the Ce signals are dependent on the orientation of the crystal inside the magnetic field, a simple rotation of the sample tube can push these signals to magnetic fields that do not interfere with the actual sample signals.

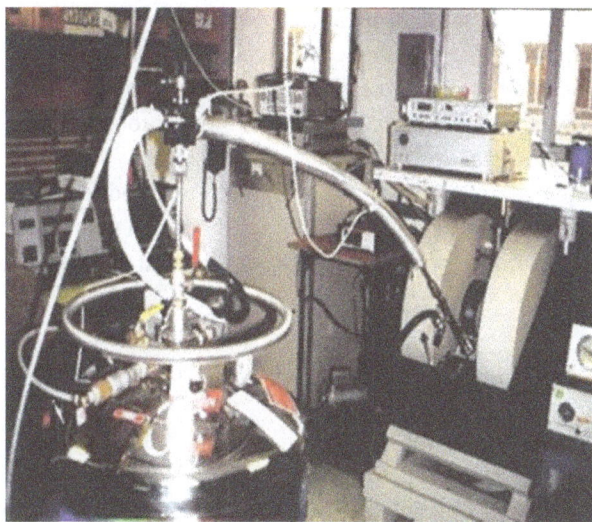

Fig. 6.18: A Bruker X-band spectrometer equipped with the Oxford Instruments ESR 900 cryostat. A flexible transfer line is employed to transfer the liquid helium from an external storage Dewar into the cryostat. (Hendrich lab, Department of Chemistry, Carnegie Mellon University)

6.3.2 Special cryogenic setups

Immersion-type cryostats are useful for many applications, especially those that require rotations of the electromagnet. Such rotations of the external magnetic field are required for single-mode resonant cavities where the magnetic field can be rotated so that it can be either perpendicular or parallel to the magnetic field component of the microwave radiation to allow observations of signals from Kramers or non-Kramers ions, respectively. These cryostats provide longer hold times for the cryogens but are more elaborate in design. In addition, the volume of samples that can be processed during a particular run is much smaller than with continuous-flow cryostats.

In the most basic immersion design, the microwave probe is immersed within a bath of liquid helium. The cavity and sample are isolated within a thin metallic can (usually made out of brass) that protects them from the liquid helium. The volume inside the metal can is pressurized with helium gas that maintains heat contact with the liquid helium bath and cools the sample down. The liquid helium Dewar is surrounded by another cylindrical Dewar which is filled with liquid nitrogen that slows down the rate of evaporation of the liquid helium. The temperature of the bath cannot be easily controlled in this type of cryostat. It can, however, be reduced to ~ 1.9 K by connecting the liquid helium bath to a mechanical pump and reducing the vapor pressure above the bath.

In a more convenient design, the microwave probe is not actually immersed in liquid helium; it is inside a Dewar that is within the liquid helium Dewar, where the liquid helium is allowed to diffuse into the sample area via a needle valve and through a diffuser. In this "immersion-type" design, the sample temperature can be more accurately controlled and can also be lowered down to 1.9 K by pumping on the liquid helium bath. This design also allows the easy removal of the microwave probe from the cryostat for sample exchange. The cryostat details are shown in Fig. 6.19.

As shown in Fig. 6.20, more elaborate microwave probes are required for these cryostats. Such probes, in addition to using special coupling methods and mounts for temperature sensors, employ safeguards to prevent excessive evaporation of cryogenic fluids.

6.4 EPR samples

EPR can be used to study a variety of samples in different forms. Crystalline, powder, liquid solution and frozen solution samples can all be studied by continuous-wave EPR. This versatility makes EPR a useful spectroscopic technique in disciplines ranging from solid-state physics and materials science to biochemistry and related biomedical areas. Crystal samples are usually preferred due to the more straightforward spectra they produce. Powder and frozen solution samples are usually more readily available since in some cases crystals of a particular molecule might be difficult to obtain. Protein samples, in particular, are typically studied in frozen solution form. Liquid samples are also used especially in the study of free radicals and enzymatic intermediates but they often present challenges when the solvent used absorbs large microwave powers.

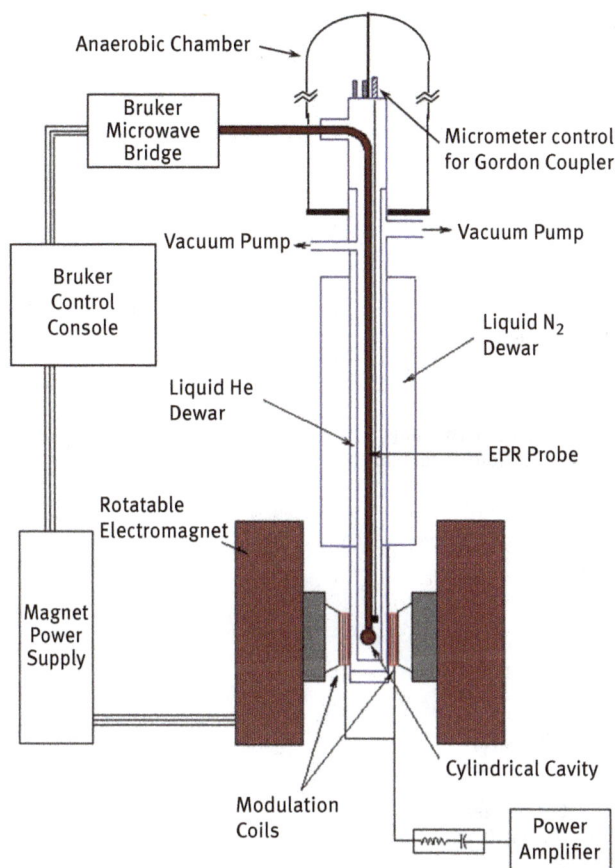

Fig. 6.19: EPR spectrometer with the microwave probe in a liquid helium immersion-type cryostat. The sample is inside the liquid He Dewar where a diffuser at the bottom allows liquid He to diffuse into the cryostat. The amount of liquid He that diffuses into the cryostat is controlled by a needle valve. Temperature sensors in the cryostat and on the microwave probe record the temperature in the cryostat and sample, respectively. This spectrometer is equipped with an anaerobic chamber for exchanging air-sensitive samples.

6.4.1 Single crystals

The EPR spectra obtained from crystalline samples are usually very strong and provide accurate information on the molecular properties of the system such as magnetic anisotropy. The size of the crystals used for EPR measurements should not be large, since even dilute crystals of a couple of millimeters in length and width can produce intense signals. The crystals require special glass holders ("EPR flats") which are cylindrical glass rods approximately 20 cm in length with a 3 mm diameter that have a 2 cm section close to one end ground to a flat to accommodate the crystal as shown

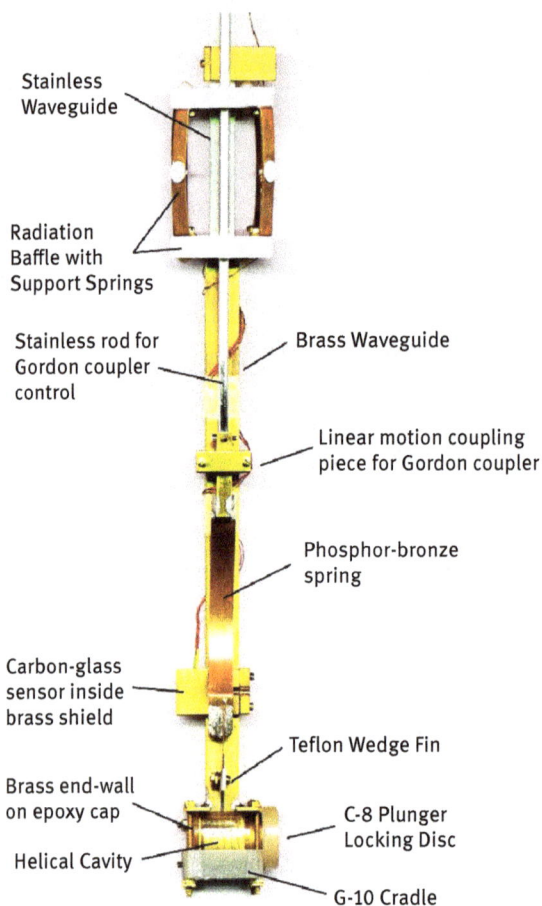

Fig. 6.20: Special-purpose microwave probe for use with the immersion-type cryostat utilized in the Q-band spectrometer shown in Fig. 6.15. This homebuilt probe employs a helical cylindrical cavity that allows orientation of the magnetic field both parallel and perpendicular to the microwave magnetic field component (see also Fig. 6.7). The cavity is coupled to the probe via a Gordon coupler. A carbo-glass temperature sensor is used to monitor the sample temperature. This sensor is inside a brass shield to prevent radio frequency heating from the modulation coils that are mounted outside the cryostat as shown in Fig. 6.15. The radiation baffle is a thermal shield that prevents room-temperature radiation from reaching the sample area.

in Fig. 6.21. A natural face of the crystal must be attached to the flat surface to ensure proper orientation. A small amount of vacuum grease, which turns into glass at cryogenic temperatures, is used to attach the crystal to the flat.

Single crystals are ideal for the study of metal–metal interactions since they can easily be synthesized with small amounts of the paramagnetic species within a diamagnetic host to reduce broadening effects due to spin–spin broadening.

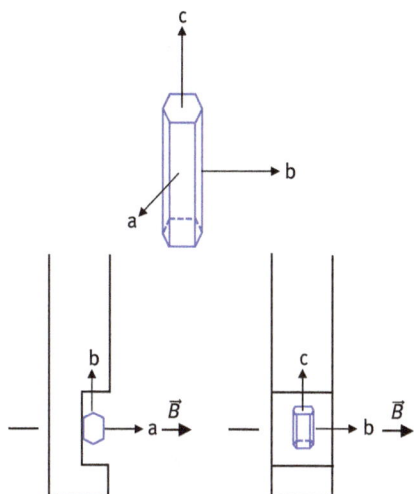

Fig. 6.21: Orienting a single crystal on an EPR flat: two perspectives of an EPR flat with two different orientations of the crystal that allow rotation of the external magnetic field to include all three crystalline axes (a, b, c). The external magnetic field rotates in a plane perpendicular to the page. The crystal habit of a hexagonal crystal with the crystal axes is shown.

6.4.2 Frozen solution samples

Frozen solution samples are widely used in the study of biological systems such as metalloproteins. Freezing such samples minimizes oxidation for air-sensitive samples and allows for long-term storage in liquid nitrogen. The solvents used must form a glass when frozen to ensure that the magnetic moments of the paramagnetic species are randomly oriented. Suitable solvents are water, glycerol, ethanol, toluene and other glassing agents. The freezing process is typically done using liquid nitrogen, where the tube with the sample and solvent is slowly lowered into the liquid nitrogen bath to ensure that the expansion of the sample occurs along the axis of the tube to prevent breakage. The slow freezing process is particularly important for aqueous samples.

EPR sample tubes are typically made from either clear fused quartz which is purified from naturally occurring quartz to minimize impurities. In some cases, tubes are made from synthetic quartz to further minimize impurities. For quantitative EPR analysis, it is essential to accurately know the cross-sectional areas of the tube which is possible with high-precision tubes.

6.4.3 Liquid samples

Liquid samples are often challenging, especially when the solvent has a large dielectric constant which interferes with the resonant absorption properties of the cavity. When such solvents interact with the electric field component $\vec{E}_1$ of the microwave radiation, they absorb large amounts of radiation resulting in large dielectric losses hat degrade the Q-factor of the cavity. This is especially true in the case of aqueous samples above 0 °C due to the large dielectric constant of liquid water. By freezing aqueous samples, we avoid this since the dielectric constant of ice is much smaller than that of water, as mentioned in the introduction to Section 6.3. To avoid large dielectric losses, the sample is placed in a position of minimum $\vec{E}_1$ as shown in Fig. 6.6 using special glass flats or capillary tubes to avoid much of the sample extending into regions of large $\vec{E}_1$. Flat cells manufactured by Wilmad-LabGlass or comparable aqueous sample holders are suitable for such samples.

References

[1] D. J. E. Ingram and J. C. Kendrew, "Electron Spin Resonance in Myoglobin and Hæmoglobin: Orientation of the Hæm Group in Myoglobin and Its Relation to the Polypeptide Chain Direction," *Nature*, vol. 178, p. 905–906, 1956.

[2] D. T. Petasis and M. P. Hendrich, "A New Q-Band EPR Probe for Quantitative Studies of Even Electron Metalloproteins," *Journal Of Magnetic Resonance*, vol. 136, no. 2, p. 200–206, 1999.

[3] C. P. Poole, Electron Spin Resonance: A Comprehensive Treatise on Experimental Techniques (2nd ed.), Mineola, New York, Dover Publications Inc., 1983.

[4] C. Bender, "Microwave Engineering Fundamentals and Spectrometer Design," In EPR Instrumental Methods, Biological Magnetic Resonance, vol. 21, C Bender and L Berliner Eds., Springer, Boston, MA, 2004.

[5] M. Hruszowiec, K. Nowak, B. Szlachetko, M. Grzelczak, W. Czarczyński, E. Pliński and T. Więckowski, "The Microwave Sources for EPR Spectroscopy," *Journal of Telecommunications and Information Technology*, no. 2, p. 18–25, 2017.

[6] T. Moreno, Microwave Transmission Design Data, New York, Dover Publications Inc., 1958.

[7] D. J. Griffiths, Introduction to Electrodynamics, Upper Saddle River, NJ, Prentice Hall, 1999.

[8] D. T. Petasis, N. S. Vander Ven, S. Simizu and S. A. Friedberg, "Pair Interactions in Pr: La (CF3SO3)3 9H2O," *Journal of Applied Physics*, vol. 73, no. 10, p. 6078–6080, 1993.

[9] Y. Sanakis, D. Petasis, V. Petrouleas and M. Hendrich, "Simultaneous Binding of Fluoride and NO to the Nonheme Iron of Photosystem II: Quantitative EPR Evidence for a Weak Exchange Interaction Between the Semiquinone QA-and the Iron-Nitrosyl Complex," *Journal of the American Chemical Society*, vol. 121, no. 39, p. 9155–9164, 1999.

[10] R. Cammack and C. E. Cooper, "Electron Paramagnetic Resonance Spectroscopy of Iron Complexes and Iron-Containing Proteins," In Methods in Enzymology: Metallobiochemistry, Part D, Physical and Spectroscopic Methods for Probing Metal Ion Environments in Metalloproteins, vol. 227, JF Riordan and BL Vllee Eds., San Diego, California, Academic Press, 1993, pp. 353–384.

[11] S. J. Campbell, I. R. Herbert, C. B. Warwick and J. M. Woodgate, "A Continuous Flow Cooling Unit for ESR Experiments Over the Temperature Range 3.7-300K," *Journal of Physics E: Scientific Instruments*, vol. 9, p. 443, 1976.

Index

https://doi.org/10.1515/9783110417562-007

www.ingramcontent.com/pod-product-compliance
Lightning Source LLC
Chambersburg PA
CBHW061409210326

41598CB00035B/6153